Studies in Computational Intelligence

Volume 703

Series editor

Janusz Kacprzyk, Polish Academy of Sciences, Warsaw, Poland
e-mail: kacprzyk@ibspan.waw.pl

About this Series

The series "Studies in Computational Intelligence" (SCI) publishes new developments and advances in the various areas of computational intelligence—quickly and with a high quality. The intent is to cover the theory, applications, and design methods of computational intelligence, as embedded in the fields of engineering, computer science, physics and life sciences, as well as the methodologies behind them. The series contains monographs, lecture notes and edited volumes in computational intelligence spanning the areas of neural networks, connectionist systems, genetic algorithms, evolutionary computation, artificial intelligence, cellular automata, self-organizing systems, soft computing, fuzzy systems, and hybrid intelligent systems. Of particular value to both the contributors and the readership are the short publication timeframe and the worldwide distribution, which enable both wide and rapid dissemination of research output.

More information about this series at http://www.springer.com/series/7092

Germano Resconi · Xiaolin Xu
Guanglin Xu

Introduction to Morphogenetic Computing

Germano Resconi
Catholic University of the Sacred Heart
Brescia
Italy

Guanglin Xu
Shanghai Lixin University of Commerce
Shanghai
China

Xiaolin Xu
Shanghai Polytechnic University
Shanghai
China

ISSN 1860-949X   ISSN 1860-9503   (electronic)
Studies in Computational Intelligence
ISBN 978-3-319-86208-8   ISBN 978-3-319-57615-2   (eBook)
DOI 10.1007/978-3-319-57615-2

Printed on acid-free paper

This Springer imprint is published by Springer Nature
The registered company is Springer International Publishing AG
The registered company address is: Gewerbestrasse 11, 6330 Cham, Switzerland

Preface

This book presents an introduction to morphogenetic computing. The idea of morphogenetic computing came from conflicts and uncertainty situations that grow up when we compare two incompatible universes as local universe and global universe, neural universe and Boolean function universe, database sink and source incompatibility fuzzy logic in the many values. For example, in recursion process we cannot find convergence, in neural network we are in local minimum, and in genetics we have a lot of instability and do not understand fully. So we must create a fundamental new approach to computation by which we can move from uncertainty, inconsistency and imprecision to a more logical stable and consistent situation. Fuzzy set, active sets and other many valued logic can be used to make reasoning in conflicts and uncertain situations but cannot reach the fundamental aim to have consistency and coherence. In morphogenetic computing, we have uncertainty which is only one step of the knowledge and the other is to establish coherent situation. Morphogenetic computing uses recursion with invariance just as in physics where experiments generate conflicts, but after we discover new models for nature where the experiments are not inconsistent but logically consistent. In this book in different situations, we show how to enter conflicts and try to escape from the conflicts and uncertain situations. We argue that global and local relation, defects in crystal non Euclidean geometry database with source and sink, genetic algorithm, neural network all become more stable and efficient when we use morphogenetic computing, where the morphogenetic means globality or morphology, field theory and other topics.

Brescia, Italy
Shanghai, China
Shanghai, China

Germano Resconi
Xiaolin Xu
Guanglin Xu

Contents

Chapter 1
Database and Graph Theory

This book presents an introduction to morphogenetic computing [27–35]. The idea of morphogenetic computing came from conflicts and uncertainty situations that grow up when we compare two incompatible universes as local universe and global universe, neural universe and Boolean function universe, database sink and source incompatibility fuzzy logic in the many values. This chapter presents the database and graph theory. Figure 1.1 is the original E-R diagram of the database including 5 entities. R is the original relation matrix representing the relations between entities [16].

Then, the relationship matrix between entities is given (as 1.1), where {Entity$_1$, Entity$_2$, ..., Entity$_n$} is set of Entities, $e_{i,j}$ is the relationship between Entity$_i$ and Entity$_j$, if there is a bi-connect of between Entity$_i$ and Entity$_j$, $e_{i,j}$ is 1, otherwise, $e_{i,j}$ is 0.

$$
\begin{bmatrix}
R & Entity_1 & Entity_2 & Entity_3 & \dots & Entity_n \\
Entity_1 & e_{1,1} & e_{1,2} & e_{1,3} & \dots & e_{1,n} \\
Entity_2 & e_{2,1} & e_{2,2} & e_{2,3} & \dots & e_{2,n} \\
Entity_3 & e_{3,1} & e_{3,2} & e_{3,3} & \dots & e_{3,n} \\
\dots & \dots & \dots & \dots & \dots & \dots \\
Entity_n & e_{n,1} & e_{n,2} & e_{n,3} & \dots & e_{n,n}
\end{bmatrix}
\tag{1.1}
$$

When we give a name to any entity we have the graph as illustrated in Fig. 1.2.

We can show the above database in matrix (oriented graph) as illustrated below 1.2.

$$
\begin{bmatrix}
R & class & classroom & enrollment & teacher & student \\
class & 0 & 1 & 1 & 1 & 0 \\
classroom & 1 & 0 & 0 & 0 & 0 \\
enrollment & 1 & 0 & 0 & 0 & 1 \\
teacher & 1 & 0 & 0 & 0 & 0 \\
student & 0 & 0 & 1 & 0 & 0
\end{bmatrix}
\tag{1.2}
$$

© Springer International Publishing AG 2017
G. Resconi et al., *Introduction to Morphogenetic Computing*,
Studies in Computational Intelligence 703, DOI 10.1007/978-3-319-57615-2_1

1

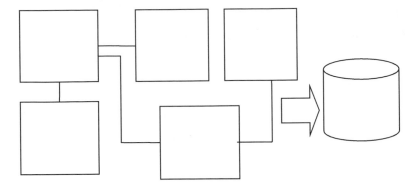

Fig. 1.1 Database scheme

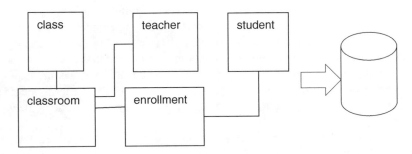

Fig. 1.2 Database with names

That can be represented as 1.3.

$$Rv = \begin{bmatrix} 0 & 1 & 1 & 1 & 0 \\ 1 & 0 & 0 & 0 & 0 \\ 1 & 0 & 0 & 0 & 1 \\ 1 & 0 & 0 & 0 & 0 \\ 0 & 0 & 1 & 0 & 0 \end{bmatrix} \begin{bmatrix} class \\ classroom \\ enrollment \\ teacher \\ student \end{bmatrix} = \begin{bmatrix} (classroom, enrollment, teacher) \\ class \\ (class, student) \\ class \\ enrollment \end{bmatrix}$$

$$(1.3)$$

The difference between 1.2 and 1.3 representation on the database in Fig. 1.2 is that the first is a static representation, the second is a dynamical representation where we can see the initial set of entities and the final set of entities. We remark that the initial set includes individual names but the final vector includes sets of entities. Any set of entities in output has one common entity with the initial entity set. Because one entity as initial value is associated with other entities, the transformation is a many value process with intrinsic uncertainty in fact from one entity we have different possible entities as final entities (bifurcation). In the database we introduce a selection rule for which in the given initial entity we select one and only one final element. The selection rule is included in the query that we write in the given initial entity.

1.1 Graph as a Space of Entity Attributes as Sink, Source, and Transit

The database graph in Fig. 1.2 can be written as a superposition of source and sink entities 1.4.

$$Rv = (Ae_1 + Be_2)v \qquad (1.4)$$

In an explicit way we have 1.5.

$$
Rv = \left(\begin{bmatrix} 1 & 0 & 0 & 0 & 0 \\ 0 & 1 & 0 & 0 & 0 \\ 0 & 0 & 1 & 0 & 0 \\ 0 & 0 & 0 & 1 & 0 \\ 0 & 0 & 0 & 0 & 1 \end{bmatrix} e_1 + \begin{bmatrix} 0 & 1 & 1 & 1 & 0 \\ 1 & 0 & 0 & 0 & 0 \\ 1 & 0 & 0 & 0 & 1 \\ 1 & 0 & 0 & 0 & 0 \\ 0 & 0 & 1 & 0 & 0 \end{bmatrix} e_2 \right) \begin{bmatrix} class \\ classroom \\ enrollment \\ teacher \\ student \end{bmatrix}
$$

$$
= \begin{bmatrix} (class) \\ (classroom) \\ (enrollment) \\ (teacher) \\ (student) \end{bmatrix} e_1 + \begin{bmatrix} (classroom, enrollment, teacher) \\ (class) \\ (class, student) \\ (class) \\ (enrollment) \end{bmatrix} e_2
$$

$$
= \begin{bmatrix} (class)e_1 + (classroom, enrollment, teacher)e_2 \\ (classroom)e_1 + (class)e_2 \\ (enrollment)e_1 + (class, student)e_2 \\ (teacher)e_1 + (class)e_2 \\ (student)e_1 + (enrollment)e_2 \end{bmatrix} \qquad (1.5)
$$

$$
= \begin{bmatrix} e_1 & e_2 & e_2 & e_2 & 0 \\ e_2 & e_1 & 0 & 0 & 0 \\ e_2 & 0 & e_1 & 0 & e_2 \\ e_2 & 0 & 0 & e_1 & 0 \\ 0 & 0 & e_2 & 0 & e_1 \end{bmatrix} \begin{bmatrix} class \\ classroom \\ enrollment \\ teacher \\ student \end{bmatrix}
$$

where e_1 are all source elements and e_2 are all sink elements for any relationship. We remark that any relationship can write as a superposition of two states one is the source state and the other is the sink state. We see that the relation R can also be written in this way 1.6.

$$R = \begin{bmatrix} e_1 & e_2 & e_2 & e_2 & 0 \\ e_2 & e_1 & 0 & 0 & 0 \\ e_2 & 0 & e_1 & 0 & e_2 \\ e_2 & 0 & 0 & e_1 & 0 \\ 0 & 0 & e_2 & 0 & e_1 \end{bmatrix} \qquad (1.6)$$

Given a row of the previous matrix 1.7.

$$R = \begin{bmatrix} e_1 & e_2 & e_2 & e_2 & 0 \\ 0 & 0 & 0 & 0 & 0 \\ 0 & 0 & 0 & 0 & 0 \\ 0 & 0 & 0 & 0 & 0 \\ 0 & 0 & 0 & 0 & 0 \end{bmatrix} \qquad (1.7)$$

This represents a divergent condition from class. In fact from the class as source we go to teacher, classroom, enrollment that are sink. The graphical representation is shown as Fig. 1.3.

Now the associate column is shown as 1.8

$$R = \begin{bmatrix} e_1 & 0 & 0 & 0 & 0 \\ e_2 & 0 & 0 & 0 & 0 \\ e_2 & 0 & 0 & 0 & 0 \\ e_2 & 0 & 0 & 0 & 0 \\ 0 & 0 & 0 & 0 & 0 \end{bmatrix} \qquad (1.8)$$

In this case any column is a convergent set of entities that are convergent into the entity class. In a graph way we have (Fig. 1.4)

From the entity "class" we have divergence and convergence, so "class" is the neutral element that includes the two states in the same entity. The row and the column has "class" as intersection so "class" belongs to the row and to the column of the matrix.

Fig. 1.3 Class as a source

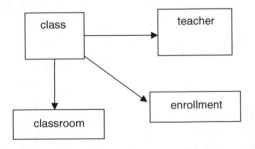

Fig. 1.4 Class as a sink

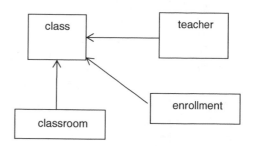

when $e_1 = 1$ and $e_2 = -1$ we have the incident matrix 1.9 that the element is negative when one node is the departure, and the element is positive when one node is the arrival.

$$R = \begin{bmatrix} -1 & 1 & 1 & 1 & 0 \\ 1 & -1 & 0 & 0 & 0 \\ 1 & 0 & -1 & 0 & 1 \\ 1 & 0 & 0 & -1 & 0 \\ 0 & 0 & 1 & 0 & -1 \end{bmatrix} \tag{1.9}$$

Now we know that the Laplacian matrix of R is 1.10.

$$L = R^T R = \begin{bmatrix} -1 & 1 & 1 & 1 & 0 \\ 1 & -1 & 0 & 0 & 0 \\ 1 & 0 & -1 & 0 & 1 \\ 1 & 0 & 0 & -1 & 0 \\ 0 & 0 & 1 & 0 & -1 \end{bmatrix}^T \begin{bmatrix} -1 & 1 & 1 & 1 & 0 \\ 1 & -1 & 0 & 0 & 0 \\ 1 & 0 & -1 & 0 & 1 \\ 1 & 0 & 0 & -1 & 0 \\ 0 & 0 & 1 & 0 & -1 \end{bmatrix}$$

$$= \begin{bmatrix} 4 & -2 & -2 & -2 & 0 \\ -2 & 2 & 1 & 1 & 0 \\ -2 & 1 & 3 & 1 & -2 \\ -2 & 1 & 1 & 2 & 0 \\ 1 & 0 & -2 & 0 & 2 \end{bmatrix}$$

$$\tag{1.10}$$

We know that Laplacian matrix gives a lot of information for the graph. Now we can extend the definition of the Laplacian matrix as 1.11.

$$L = R^T R = \begin{bmatrix} e_1 & e_2 & e_2 & e_2 & 0 \\ e_2 & e_1 & 0 & 0 & 0 \\ e_2 & 0 & e_1 & 0 & e_2 \\ e_2 & 0 & 0 & e_1 & 0 \\ 0 & 0 & e_2 & 0 & e_1 \end{bmatrix}^T \begin{bmatrix} e_1 & e_2 & e_2 & e_2 & 0 \\ e_2 & e_1 & 0 & 0 & 0 \\ e_2 & 0 & e_1 & 0 & e_2 \\ e_2 & 0 & 0 & e_1 & 0 \\ 0 & 0 & e_2 & 0 & e_1 \end{bmatrix}$$

$$= \begin{bmatrix} e_1^2 + 3e_2^2 & e_1e_2 + e_2e_1 & e_1e_2 + e_2e_1 & e_1e_2 + e_2e_1 & e_2^2 \\ e_1e_2 + e_2e_1 & e_1^2 + e_2^2 & e_2^2 & e_2^2 & 0 \\ e_1e_2 + e_2e_1 & e_2^2 & 2e_1^2 + e_2^2 & e_2^2 & e_1e_2 + e_2e_1 \\ e_1e_2 + e_2e_1 & e_2^2 & e_2^2 & e_1^2 + e_2^2 & 0 \\ e_2^2 & 0 & e_1e_2 + e_2e_1 & 0 & e_1^2 + e_2^2 \end{bmatrix}$$

$$(1.11)$$

Because we have the transport matrix we multiply the convergent part of one entity with the convergent part of another entity and superpose the convergent parts. For example the multiplication of the convergent part of the same entity "class" we have 1.12.

$$\begin{bmatrix} class \\ e_1 \\ e_2 \\ e_2 \\ e_2 \\ 0 \end{bmatrix} \begin{bmatrix} class \\ e_1 \\ e_2 \\ e_2 \\ e_2 \\ 0 \end{bmatrix} = \begin{bmatrix} class^2 \\ e_1^2 \\ e_2^2 \\ e_2^2 \\ e_2^2 \\ 0 \end{bmatrix} \qquad (1.12)$$

And the superposition of the convergent parts is 1.13.

$$e_1^2 + e_2^2 + e_2^2 + e_2^2 = e_1^2 + 3e_2^2 \qquad (1.13)$$

Given two different entities we have the product of the convergent parts. The first column are entities that converge to class from classroom, teacher and enrollment, the second column are entities that converge on the classroom that is only one from class. The products are common links that converge to class and that converge to classroom.

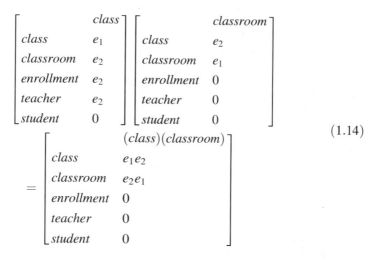

$$(1.14)$$

The column "class" has the set of ordered convergent elements.

$$class = (class, classroom, enrollment, teacher)$$

The column "classroom" has the set of ordered convergent elements.

$$classroom = (class, classroom)$$

The ordered intersection is

$$(class)(classroom) = ((class)(class) = e_1e_2, (classroom)(classroom) = e_2e_1)$$

The product of class and classroom give us all the common convergent links including the state e_1 that is the neutral element and is divergent and convergent in class and classroom. In a graph way we have Fig. 1.5.

Fig. 1.5 The product of class and classroom

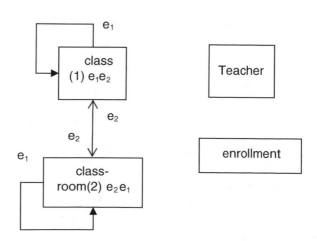

To explain in more specific way, we can decompose relation R as 1.15.

$$
\begin{bmatrix}
R_1 & class & classroom & enrollmnet & teacher & student \\
class & 0 & 1 & 1 & 1 & 0 \\
classroom & 0 & 0 & 0 & 0 & 0 \\
enrollment & 0 & 0 & 0 & 0 & 0 \\
teacher & 0 & 0 & 0 & 0 & 0 \\
student & 0 & 0 & 0 & 0 & 0
\end{bmatrix}
$$

$$
+
\begin{bmatrix}
R_2 & class & classroom & enrollmnet & teacher & student \\
class & 0 & 0 & 0 & 0 & 0 \\
classroom & 1 & 0 & 0 & 0 & 0 \\
enrollment & 0 & 0 & 0 & 0 & 0 \\
teacher & 0 & 0 & 0 & 0 & 0 \\
student & 0 & 0 & 0 & 0 & 0
\end{bmatrix}
$$

$$
+
\begin{bmatrix}
R_3 & class & classroom & enrollmnet & teacher & student \\
class & 0 & 0 & 0 & 0 & 0 \\
classroom & 0 & 0 & 0 & 0 & 0 \\
enrollment & 1 & 0 & 0 & 0 & 1 \\
teacher & 0 & 0 & 0 & 0 & 0 \\
student & 0 & 0 & 0 & 0 & 0
\end{bmatrix}
\quad (1.15)
$$

$$
+
\begin{bmatrix}
R_4 & class & classroom & enrollmnet & teacher & student \\
class & 0 & 0 & 0 & 0 & 0 \\
classroom & 0 & 0 & 0 & 0 & 0 \\
enrollment & 0 & 0 & 0 & 0 & 0 \\
teacher & 1 & 0 & 0 & 0 & 0 \\
student & 0 & 0 & 0 & 0 & 0
\end{bmatrix}
$$

$$
+
\begin{bmatrix}
R_5 & class & classroom & enrollmnet & teacher & student \\
class & 0 & 0 & 0 & 0 & 0 \\
classroom & 0 & 0 & 0 & 0 & 0 \\
enrollment & 0 & 0 & 0 & 0 & 0 \\
teacher & 0 & 0 & 0 & 0 & 0 \\
student & 1 & 0 & 1 & 0 & 0
\end{bmatrix}
$$

For the five relations we can see that given an entity whose column values are all zero, that means no any other entity in this relation has access to the entity but from the entity in the row we have many other entities so the entity is a source but is not a sink. Reversely we have the entity whose row values are all zero but the column values are not. That means other entities in this relation have access to the entity but there is no relation from the entity so in this situation the entity is a sink. For example in R_1 class is a source, and classroom, enrollment and teacher are sinks. Because we have only one row different from zero values, we have only one source at the time or e_1 and many different sinks or e_2. Now we also introduce the transit. We can see in this decomposition 1.16.

$$
\begin{bmatrix}
R_{12} & class & classroom & enrollmnet & teacher & student \\
class & 0 & 1 & 1 & 1 & 0 \\
classroom & 1 & 0 & 0 & 0 & 0 \\
enrollment & 0 & 0 & 0 & 0 & 0 \\
teacher & 0 & 0 & 0 & 0 & 0 \\
student & 0 & 0 & 0 & 0 & 0
\end{bmatrix}
\tag{1.16}
$$

In the previous decomposition we have that classroom has the column and row with values not all zero. So classroom is a transit element that connects class with itself. In fact from class as a source we go to classroom as a sink but classroom is also a source that go to the class as a final sink. Now we have that R_{12} can write as 1.17.

$$
\begin{bmatrix}
R_{12} & class & classroom & enrollmnet & teacher & student \\
class & 0 & 1 & 1 & 1 & 0 \\
classroom & 1 & 0 & 0 & 0 & 0 \\
enrollment & 0 & 0 & 0 & 0 & 0 \\
teacher & 0 & 0 & 0 & 0 & 0 \\
student & 0 & 0 & 0 & 0 & 0
\end{bmatrix}
=
$$

$$
\begin{bmatrix}
R_1 & class & classroom & enrollmnet & teacher & student \\
class & 0 & 1 & 1 & 1 & 0 \\
classroom & 0 & 0 & 0 & 0 & 0 \\
enrollment & 0 & 0 & 0 & 0 & 0 \\
teacher & 0 & 0 & 0 & 0 & 0 \\
student & 0 & 0 & 0 & 0 & 0
\end{bmatrix}
\tag{1.17}
$$

$$
+
\begin{bmatrix}
R_2 & class & classroom & enrollmnet & teacher & student \\
class & 0 & 0 & 0 & 0 & 0 \\
classroom & 1 & 0 & 0 & 0 & 0 \\
enrollment & 0 & 0 & 0 & 0 & 0 \\
teacher & 0 & 0 & 0 & 0 & 0 \\
student & 0 & 0 & 0 & 0 & 0
\end{bmatrix}
$$

For a path that moves from class to class room and from class room go to class again, we have that the path is a superposition of four states. The first state is the source for the first link, the second state is the sink for the first link, the third state is the source for the second link ad the fourth state is the sink for the second link. The two links are superpose in the path. In fact we have

$$(class, classroom) \underset{e_1}{\rightarrow} \underset{e_2}{\rightarrow} (classroom, class)$$

where

$$Link_1 = (class, classroom) = class \rightarrow classroom$$
$$Link_2 = (classroom, class) = classroom \rightarrow clas$$

So

$$(class, classroom)e_1 + (classroom, class)e_2 = (Link_1)e_1 + (Link_2)e_2$$
and
$$Link_1 = (class)e_1 + (classroom)e_2$$
$$Link_2 = (classroom)e_1 + (class)e_2$$

and

$$(class)e_1e_1 + (classroom)e_2e_1 + (classroom)e_1e_2 + (class)e_2e_2$$

That can be written in this simple way.

$$R_1R_2 = (class)e_{11} + (classroom)e_{21} + (classroom)e_{12} + (class)e_{22}$$

where e_{11} is the first initial value for the first step, e_{21} is the final value for the first step, e_{12} is the initial value for the second step and e_{22} is the final value for the second step. In conclusion, to represent two joined steps we use four dimensions space, and for only one step we use two dimensions.

When we join source and sink with many different transit entities, we move from two dimensions to four, eight and so on dimensions.

1.2   Derivative, Variation and Chain by the Reference e_1 and e_2

Given the chain

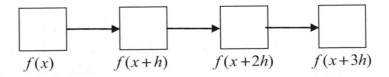

$$f(x) \qquad f(x+h) \qquad f(x+2h) \qquad f(x+3h)$$

Difference when $e_1 = -1$, $e_2 = 1$

$$f(x)e_1 + f(x+h)e_2 = f(x+h)e_2 - f(x)e_1 = \Delta f = direction$$
$$(f(x)e_1 + f(x+h)e_2)e_1 + (f(x+h)e_1 + f(x+2h)e_2)e_2$$
$$= (f(x+h) - f(x))e_1 + (f(x+2h) - f(x+h))e_2$$
$$= -(f(x+h) - f(x)) + (f(x+2h) - f(x+h)) = \Delta^2 f = curvature$$
$$[(f(x+h) - f(x))e_1 + (f(x+2h) - f(x+h))e_2]e_1$$
$$+ [(f(x+2h) - f(x+h))e_1 + (f(x+3h) - f(x+2h))e_2]e_2 = \Delta^3 f = Torsion$$

Variation of the product by the reference e_1 and e_2

$$f(x)g(x)e_1 + f(x+h)g(x+h)e_2$$
$$= f(x)g(x)e_1 + f(x+h)g(x+h)e_2 + f(x)g(x+h)(e_1 + e_2)$$
$$= f(x)(g(x)e_1 + g(x+h)e_2) + g(x+h)(f(x)e_1 + f(x+h)e_2)$$

We remember that

$$f(x)g(x+h)(e_1 + e_2) = 0$$

Chapter 2
Crossover and Permutation

Given the permutation P

$$P = \begin{pmatrix} 1 & 2 & 3 & \ldots & n-1 & n \\ p_1 & p_2 & p_3 & \ldots & p_{n-1} & p_n \end{pmatrix}$$

Given the two crossovers

$$U = \begin{bmatrix} & A & B \\ C & a & a \\ D & b & b \end{bmatrix} \rightarrow \begin{bmatrix} DA & CB \\ b & a \end{bmatrix}$$

$$U = \begin{bmatrix} & A & B \\ C & a & a \\ D & b & b \end{bmatrix} \rightarrow \begin{bmatrix} CA & DB \\ a & b \end{bmatrix}$$

We have the elementary permutation

$$P = \begin{pmatrix} a & b \\ b & a \end{pmatrix}$$

For more simple crossover we can create the matrix M

$$M = \begin{pmatrix} a & a \\ b & b \end{pmatrix}$$

© Springer International Publishing AG 2017
G. Resconi et al., *Introduction to Morphogenetic Computing*,
Studies in Computational Intelligence 703, DOI 10.1007/978-3-319-57615-2_2

where we have two parents and two genes so we have the crossover ab, ba that are the terms of the permutation. For three parents and three genes we have the six possible crossovers from the three parents.

$$M = \begin{bmatrix} a & a & a \\ b & b & b \\ c & c & c \end{bmatrix} \rightarrow abc, acb, bac, bca, cab, cba$$

The permutation matrix is

$$a_{h,k} = \delta_{h,p_k}, \quad where \quad \begin{cases} \delta_{k,p_k} = 1 \\ \delta_{h,p_k} = 0, \quad h \neq k \end{cases}$$

For example, given the permutation P,

$$P = \begin{pmatrix} 1 & 2 & 3 \\ p_1 = 2 & p_2 = 1 & p_3 = 3 \end{pmatrix}$$

the permutation matrix is A.

$$A = \begin{bmatrix} & p_1 & p_2 & p_3 \\ k_1 & 0 & 1 & 0 \\ k_2 & 1 & 0 & 0 \\ k_3 & 0 & 0 & 1 \end{bmatrix} \rightarrow \begin{bmatrix} 0 & 1 & 0 \\ 1 & 0 & 0 \\ 0 & 0 & 1 \end{bmatrix}$$

With the permutation matrix we permute the columns by multiplication of R at the right. We have

$$RA = \begin{bmatrix} e_{11} & e_{12} & e_{13} \\ e_{21} & e_{22} & e_{23} \\ e_{31} & e_{32} & e_{33} \end{bmatrix} \begin{bmatrix} 0 & 1 & 0 \\ 1 & 0 & 0 \\ 0 & 0 & 1 \end{bmatrix} = \begin{bmatrix} e_{12} & e_{11} & e_{13} \\ e_{22} & e_{21} & e_{23} \\ e_{32} & e_{31} & e_{33} \end{bmatrix}$$

So we get RA by right multiplication R with permutation matrix A. And we get AR by left multiplication R with permutation matrix A.

$$AR = \begin{bmatrix} 0 & 1 & 0 \\ 1 & 0 & 0 \\ 0 & 0 & 1 \end{bmatrix} \begin{bmatrix} e_{11} & e_{12} & e_{13} \\ e_{21} & e_{22} & e_{23} \\ e_{31} & e_{32} & e_{33} \end{bmatrix} = \begin{bmatrix} e_{21} & e_{22} & e_{23} \\ e_{11} & e_{12} & e_{13} \\ e_{31} & e_{32} & e_{33} \end{bmatrix}$$

What is the difference between RA and AR?

In RA, Column 1 and Column 2 of R are swapped. Whereas in AR, Row 1 and Row 2 are swapped.

We write the difference between RA and AR as 2.1.

$$
\begin{bmatrix} 0 & 1 & 0 \\ 1 & 0 & 0 \\ 0 & 0 & 1 \end{bmatrix} \begin{bmatrix} e_{11} & e_{12} & e_{13} \\ e_{21} & e_{22} & e_{23} \\ e_{31} & e_{32} & e_{33} \end{bmatrix} - \begin{bmatrix} e_{11} & e_{12} & e_{13} \\ e_{21} & e_{22} & e_{23} \\ e_{31} & e_{32} & e_{33} \end{bmatrix} \begin{bmatrix} 0 & 1 & 0 \\ 1 & 0 & 0 \\ 0 & 0 & 1 \end{bmatrix}
$$

$$
= \begin{bmatrix} e_{21} & e_{22} & e_{23} \\ e_{31} & e_{32} & e_{33} \\ e_{31} & e_{32} & e_{33} \end{bmatrix} - \begin{bmatrix} e_{12} & e_{11} & e_{13} \\ e_{22} & e_{21} & e_{23} \\ e_{32} & e_{31} & e_{33} \end{bmatrix} \tag{2.1}
$$

$$
= \begin{bmatrix} e_{21} - e_{12} & e_{22} - e_{11} & e_{23} - e_{13} \\ e_{31} - e_{21} & e_{32} - e_{22} & e_{33} - e_{23} \\ e_{31} - e_{31} & e_{32} - e_{31} & 0 \end{bmatrix}
$$

2.1   Right Product RA

Given Relation R and permutation matrix A, we have RA (2.2).

$$
R = \begin{bmatrix} 0 & 1 & 1 & 1 & 0 \\ 1 & 0 & 0 & 0 & 0 \\ 1 & 0 & 0 & 0 & 1 \\ 1 & 0 & 0 & 0 & 0 \\ 0 & 0 & 1 & 0 & 0 \end{bmatrix} \quad A = \begin{bmatrix} 0 & 1 & 0 & 0 & 0 \\ 1 & 0 & 0 & 0 & 0 \\ 0 & 0 & 1 & 0 & 0 \\ 0 & 0 & 0 & 1 & 0 \\ 0 & 0 & 0 & 0 & 1 \end{bmatrix}
$$

$$
RA = \begin{bmatrix} 1 & 0 & 1 & 1 & 0 \\ 0 & 1 & 0 & 0 & 0 \\ 0 & 1 & 0 & 0 & 1 \\ 0 & 1 & 0 & 0 & 0 \\ 0 & 0 & 1 & 0 & 0 \end{bmatrix} \tag{2.2}
$$

$$
\begin{bmatrix} 1 & 0 & 1 & 1 & 0 \\ 0 & 1 & 0 & 0 & 0 \\ 0 & 1 & 0 & 0 & 1 \\ 0 & 1 & 0 & 0 & 0 \\ 0 & 0 & 1 & 0 & 0 \end{bmatrix} \begin{bmatrix} class \\ classroom \\ enrollment \\ teacher \\ student \end{bmatrix} \rightarrow \begin{bmatrix} class + enrollment + teacher \\ classroom \\ classroom + student \\ classroom \\ enrollment \end{bmatrix}
$$

And the relation in the database of Fig. 1.2 becomes that in Fig. 2.1.

In Fig. 2.1, the changes of relations just happen on the entities directly related to entity class (represented as 1) and entity classroom (represented as 2), the reason for this is because multiplying permutation matrix A at the right of R makes R the first two columns swap. This leads to the disappearance of some relations and the generation of some new relations. For example, originally there is the relation from enrollment to class (from 3 to 1) and there is no relation from enrollment to

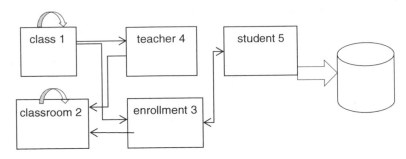

Fig. 2.1 The database scheme with right product

classroom (from 3 to 2), after permutation, the relation from enrollment to class (from 3 to 1) disappears and relation from enrollment to classroom (from 3 to 2) emerges. From the diagram, we also find the relations between 3 and 5, and those between 4 and 5 don't change on the grounds that relations between them have nothing to do with entity 1 and entity 2. Further it is not difficult to find that the number of relations doesn't change. The reason for this is that we just swap the two columns of Matrix R, not cause any change on the number of 1 in the relation RA. So, from the permutation above, what conclusion can we make? In RA we have that $e_{i,j} \rightarrow e_{i,p_j}$ or in a graphic way we have Fig. 2.2.

Here

$$\begin{bmatrix} k & 1 & 2 & 3 \\ p_k & 2 & 1 & 3 \end{bmatrix}$$

In the right product RA the initial entity is the same, but the final element changes for the permutation.

We give the relations change of database.

We know that teacher has access to class but at one time he wants to have access to another entity as classroom so we permutate class with classroom and we use the permutation A in a way to change all the other entities to satisfy teacher without changing the number of relations. So we have Fig. 2.3.

Fig. 2.2 Right product RA

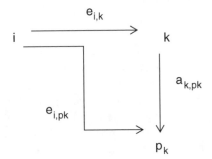

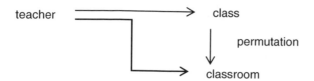

Fig. 2.3 Permutate class with classroom by right product

2.2 Left Product AR

The product AR changes the rows so we have 2.3

$$AR = \begin{bmatrix} 1 & 0 & 0 & 0 & 0 \\ 0 & 1 & 1 & 1 & 0 \\ 1 & 0 & 0 & 0 & 1 \\ 1 & 0 & 0 & 0 & 0 \\ 0 & 0 & 1 & 0 & 0 \end{bmatrix} \tag{2.3}$$

$$\begin{bmatrix} 1 & 0 & 0 & 0 & 0 \\ 0 & 1 & 1 & 1 & 0 \\ 1 & 0 & 0 & 0 & 1 \\ 1 & 0 & 0 & 0 & 0 \\ 0 & 0 & 1 & 0 & 0 \end{bmatrix} \begin{bmatrix} class \\ classroom \\ enrollment \\ teacher \\ student \end{bmatrix} \rightarrow \begin{bmatrix} class \\ classroom + enrollment + teacher \\ class + student \\ class \\ enrollment \end{bmatrix}$$

So the relation in the database of Fig. 1.2 becomes that in Fig. 2.4.

In AR, the final entity is the same, but the initial element changes for the permutation.

We give the relations change of database.

Initially we can find the relevant teacher via class. Now on some occasions that we are urged to find the teacher, going to the relevant classroom is the only way. So by permutation, we get the relation between classroom and teacher. So we have Fig. 2.5.

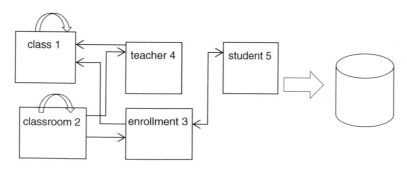

Fig. 2.4 The database scheme with left product

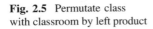

Fig. 2.5 Permutate class
with classroom by left product

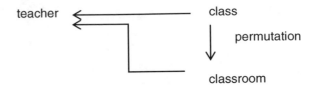

When we permute two entities by some way, all relations related to the two entities change correspondingly, although some relations change may not be necessary. So the graph will change.

Chapter 3
Similarity Between Graphs in Database by Permutations

Given the relationship matrix R (3.1) that reflects the connection among entities.

$$
\begin{bmatrix}
R & Entity_1 & Entity_2 & Entity_3 & \cdots & Entity_n \\
Entity_1 & e_{1,1} & e_{1,2} & e_{1,3} & \cdots & e_{1,n} \\
Entity_2 & e_{2,1} & e_{2,2} & e_{2,3} & \cdots & e_{2,n} \\
Entity_3 & e_{3,1} & e_{3,2} & e_{3,3} & \cdots & e_{3,n} \\
\cdots & \cdots & \cdots & \cdots & \cdots & \cdots \\
Entity_n & e_{n,1} & e_{n,2} & e_{n,3} & \cdots & e_{n,n}
\end{bmatrix}
\tag{3.1}
$$

Then permutation matrix A has been created. With two relations R and two permutations, we can built the commutative diagram Fig. 3.1.

We remark that when $R_1 A_1 = A_2 R_2$ the diagram becomes a commutative diagram for which the relations of similarity between two databases is shown as 3.2.

$$
R_1 = A_2 R_2 A_1^{-1}
\tag{3.2}
$$

We remember that permutation matrix has always the inverse, so we have 3.3.

$$
R_2 = A_2^{-1} R_1 A_1
\tag{3.3}
$$

Example 3.1 We have the Relation Fig. 3.2.

We represent it with R_2.

$$
R_2 = \left(
\begin{bmatrix}
1 & 0 & 0 & 0 & 0 \\
0 & 1 & 0 & 0 & 0 \\
0 & 0 & 1 & 0 & 0 \\
0 & 0 & 0 & 1 & 0 \\
0 & 0 & 0 & 0 & 1
\end{bmatrix} e_1 +
\begin{bmatrix}
0 & 1 & 1 & 1 & 0 \\
1 & 0 & 0 & 0 & 0 \\
1 & 0 & 0 & 0 & 1 \\
1 & 0 & 0 & 0 & 0 \\
0 & 0 & 1 & 0 & 0
\end{bmatrix} e_2
\right)
$$

© Springer International Publishing AG 2017
G. Resconi et al., *Introduction to Morphogenetic Computing*,
Studies in Computational Intelligence 703, DOI 10.1007/978-3-319-57615-2_3

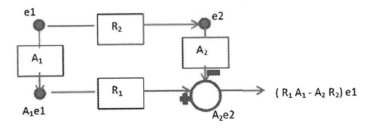

Fig. 3.1 The change of sources and sinks to give equivalent graphs

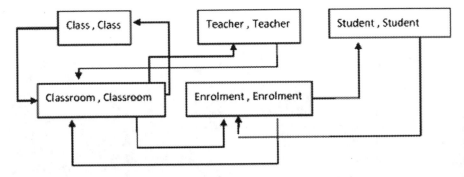

Fig. 3.2 Sources and sinks in the database

After R_2 is given the permutation with permutation matrix A_1 and A_2, we have R_1.

$$A_1 = \begin{bmatrix} 0 & 1 & 0 & 0 & 0 \\ 1 & 0 & 0 & 0 & 0 \\ 0 & 0 & 1 & 0 & 0 \\ 0 & 0 & 0 & 1 & 0 \\ 0 & 0 & 0 & 0 & 1 \end{bmatrix}, \quad A_2 = \begin{bmatrix} 1 & 0 & 0 & 0 & 0 \\ 0 & 1 & 0 & 0 & 0 \\ 0 & 0 & 1 & 0 & 0 \\ 0 & 0 & 0 & 0 & 1 \\ 0 & 0 & 0 & 1 & 0 \end{bmatrix}$$

$$R_1 = \left(\begin{bmatrix} 1 & 0 & 0 & 0 & 0 \\ 0 & 1 & 0 & 0 & 0 \\ 0 & 0 & 1 & 0 & 0 \\ 0 & 0 & 0 & 1 & 0 \\ 0 & 0 & 0 & 0 & 1 \end{bmatrix} A_1 e_1 + \begin{bmatrix} 0 & 1 & 1 & 1 & 0 \\ 1 & 0 & 0 & 0 & 0 \\ 1 & 0 & 0 & 0 & 1 \\ 1 & 0 & 0 & 0 & 0 \\ 0 & 0 & 1 & 0 & 0 \end{bmatrix} A_2 e_2 \right)$$

The relation R_1 is represented with Fig. 3.3.
Particular case for the diagram.

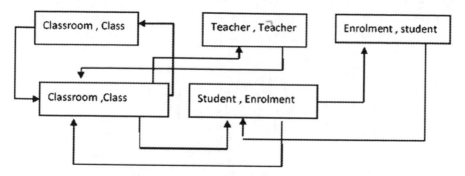

Fig. 3.3 Permutation of the sources states e_1 and permutation of the sinks states e_2 in the database

(1) $A_1 = A_2 = A$

$$R_1 = \left(\begin{bmatrix} 1 & 0 & 0 & 0 & 0 \\ 0 & 1 & 0 & 0 & 0 \\ 0 & 0 & 1 & 0 & 0 \\ 0 & 0 & 0 & 1 & 0 \\ 0 & 0 & 0 & 0 & 1 \end{bmatrix} Ae_1 + \begin{bmatrix} 0 & 1 & 1 & 1 & 0 \\ 1 & 0 & 0 & 0 & 0 \\ 1 & 0 & 0 & 0 & 1 \\ 1 & 0 & 0 & 0 & 0 \\ 0 & 0 & 1 & 0 & 0 \end{bmatrix} Ae_2 \right)$$

$$= A \left(\begin{bmatrix} 1 & 0 & 0 & 0 & 0 \\ 0 & 1 & 0 & 0 & 0 \\ 0 & 0 & 1 & 0 & 0 \\ 0 & 0 & 0 & 1 & 0 \\ 0 & 0 & 0 & 0 & 1 \end{bmatrix} e_1 + \begin{bmatrix} 0 & 1 & 1 & 1 & 0 \\ 1 & 0 & 0 & 0 & 0 \\ 1 & 0 & 0 & 0 & 1 \\ 1 & 0 & 0 & 0 & 0 \\ 0 & 0 & 1 & 0 & 0 \end{bmatrix} e_2 \right)$$

For

$$A = \begin{bmatrix} 0 & 1 & 0 & 0 & 0 \\ 1 & 0 & 0 & 0 & 0 \\ 0 & 0 & 1 & 0 & 0 \\ 0 & 0 & 0 & 1 & 0 \\ 0 & 0 & 0 & 0 & 1 \end{bmatrix}$$

(2) $A_1 \neq A_2, \quad R_1 = R_2$

For

$$A_1 = \begin{bmatrix} 0 & 1 & 0 & 0 & 0 \\ 1 & 1 & 0 & 0 & 0 \\ 0 & 0 & 1 & 0 & 0 \\ 0 & 0 & 0 & 1 & 0 \\ 0 & 0 & 0 & 0 & 1 \end{bmatrix}$$

We have

$$A_2 = \left(\begin{bmatrix} 1 & 0 & 0 & 0 & 0 \\ 0 & 1 & 0 & 0 & 0 \\ 0 & 0 & 1 & 0 & 0 \\ 0 & 0 & 0 & 1 & 0 \\ 0 & 0 & 0 & 0 & 1 \end{bmatrix} Re_1 + \begin{bmatrix} 0 & 1 & 0 & 0 & 0 \\ 1 & 0 & 0 & 0 & 0 \\ 0 & 0 & 1 & 0 & 0 \\ 0 & 0 & 0 & 1 & 0 \\ 0 & 0 & 0 & 0 & 1 \end{bmatrix} Re_2 \right)$$

$$= R \left(\begin{bmatrix} 1 & 0 & 0 & 0 & 0 \\ 0 & 1 & 0 & 0 & 0 \\ 0 & 0 & 1 & 0 & 0 \\ 0 & 0 & 0 & 1 & 0 \\ 0 & 0 & 0 & 0 & 1 \end{bmatrix} e_1 + \begin{bmatrix} 0 & 1 & 0 & 0 & 0 \\ 1 & 1 & 0 & 0 & 0 \\ 0 & 0 & 1 & 0 & 0 \\ 0 & 0 & 0 & 1 & 0 \\ 0 & 0 & 0 & 0 & 1 \end{bmatrix} e_2 \right)$$

For $R = \begin{bmatrix} 0 & 1 & 1 & 1 & 0 \\ 1 & 0 & 0 & 0 & 0 \\ 1 & 0 & 0 & 0 & 1 \\ 1 & 0 & 0 & 0 & 0 \\ 0 & 0 & 1 & 0 & 0 \end{bmatrix}$, We have

$$A_2 = \begin{bmatrix} 0 & 1 & 1 & 1 & 0 \\ 1 & 0 & 0 & 0 & 0 \\ 1 & 0 & 0 & 0 & 1 \\ 1 & 0 & 0 & 0 & 0 \\ 0 & 0 & 1 & 0 & 0 \end{bmatrix} \left(\begin{bmatrix} 1 & 0 & 0 & 0 & 0 \\ 0 & 1 & 0 & 0 & 0 \\ 0 & 0 & 1 & 0 & 0 \\ 0 & 0 & 0 & 1 & 0 \\ 0 & 0 & 0 & 0 & 1 \end{bmatrix} e_1 + \begin{bmatrix} 0 & 1 & 0 & 0 & 0 \\ 1 & 0 & 0 & 0 & 0 \\ 0 & 0 & 1 & 0 & 0 \\ 0 & 0 & 0 & 1 & 0 \\ 0 & 0 & 0 & 0 & 1 \end{bmatrix} e_2 \right)$$

$$= \begin{bmatrix} 0 & 1 & 1 & 1 & 0 \\ 1 & 0 & 0 & 0 & 0 \\ 1 & 0 & 0 & 0 & 1 \\ 1 & 0 & 0 & 0 & 0 \\ 0 & 0 & 1 & 0 & 0 \end{bmatrix} e_1 + \begin{bmatrix} 1 & 0 & 1 & 1 & 0 \\ 0 & 1 & 1 & 1 & 0 \\ 0 & 1 & 0 & 0 & 1 \\ 0 & 1 & 0 & 0 & 0 \\ 0 & 0 & 1 & 0 & 0 \end{bmatrix} e_2$$

Chapter 4
Morphogenetic and Morpheme Network to Structured Worlds

4.1 Morpheme Networks

In Fig. 4.1 we show the chaotic structure of the language before the building of the network of morphology.

Morphemes are the smallest meaningful parts of words and therefore represent a natural unit to study the evolution of words. Using a network approach from bioinformatics, we examine the historical dynamics of morphemes, the fixation of new morphemes and the emergence of words containing existing morphemes. We find that these processes are driven mainly by the number of different direct neighbors of a morpheme in words (connectivity, an equivalent to family size or type frequency) and not its frequency of usage (equivalent to token frequency).

As morphemes are also relevant for the mental representation of words, this result might enable to establish a link between an individual's perception of language and historical language change. Methods developed for the study of biological evolution might be useful for the analysis of language change.

The factors driving language change can be classified as internal and external ones. The internal factors are the physical conditions, like the physiology of the human speech organs and psychological factors like perception, processing and learning of language. On the other hand, the external factors are for example expressive use, prestige and stigma, education and language contact. In the case of words it was shown quantitatively, that the parts which compose a word. So called morphemes are the minimal meaning bearing units of words. As one word can be built by multiple morphemes, one morpheme can be found in different words. The study of how these morphemes can be combined to yield words is the central question of morphology. In this descriptive structural linguistic view, morphemes are seen as discrete units which are combined to build words. There has connectionists approaches assume that 'the same general principles that govern phonological and semantic processing of whole words and sentences govern the processing of the subparts of words commonly called morphemes of such residual

© Springer International Publishing AG 2017
G. Resconi et al., *Introduction to Morphogenetic Computing*,
Studies in Computational Intelligence 703, DOI 10.1007/978-3-319-57615-2_4

Fig. 4.1 Chaotic unstructured language

effects by exploiting an analogy of words and proteins which enabled the application of an approach from bioinformatics. Usually, arguments in favor of one or the other model are drawn from psycholinguistic studies of well selected small sets of words.

For analyzing the morphemes and their relationships, we used an approach which was successfully applied to the analysis of proteins and domains, the structural, functional and evolutionary units of proteins. Like a morpheme in words, one domain can be found in different proteins and one protein can harbor many domains. We used this analogy to build morpheme networks. Here, morphemes are nodes which are connected if they can be found next to each other in at least one word, see Fig. 4.2.

Figure 4.3 illustrates the lexical network. Perceived pictures (e.g., of a dog) directly activate concept nodes and perceived words (e.g., DOG) directly activate

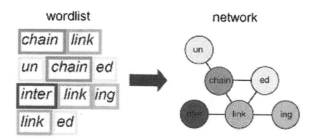

Fig. 4.2 Morpheme network

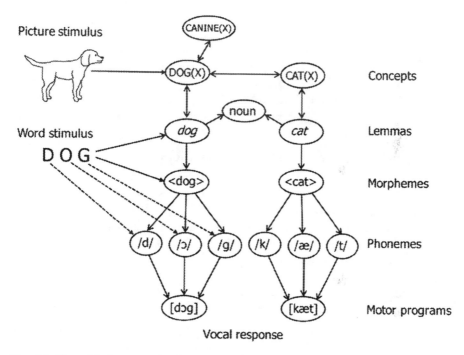

Fig. 4.3 Hierarchical structure for the word DOG

lemma, morpheme, and phoneme nodes, after which other nodes become activated through spreading activation. The dashed lines indicate grapheme-to-phoneme correspondences. Thus, our focus is on formatives, which 'recur in the morphological analysis of word-forms' independent of whether or not they are also morphemes. This fits well to the algorithm implemented by Morfessor 1.0, which searches for the optimal concise set of units Connectivity, Not Frequency, Determines the Fate of a Morpheme every word in the data can be formed by concatenation of some unit A network was built for each word list with morphemes as nodes and an undirected edge between morphemes if they occur side by side in a word. Thus, when analyzing the word 'beautifulness', no edge between 'beauti' and 'ness' would be drawn, as these are no direct neighbors. Analyses with directed edges (according to reading order) gave similar results.

4.2   Loop and General Similarity and Conflicts and Inconsistency in Graph Space

The system axiom in A. Wayne Wymore is that any system element or entity has one name. In cloud computing with uncertainty, any entity can have two or more conflicting names. At any entity we associate n ports in input and n ports in output with different names. All ports in a network is represented by a matrix in which for any row we have all names in input and output for one entity and for any column we have the names in the same states for all the entities. Morphogenetic of a network by cross over transformations of a prototype permutation (crossover) invariants (virus as a chemical network).

4.3   Vector Representation of Graph Inconsistency

Morphic has two basic elements, one is relationship, and the other is node. Additionally, there is no order in the Morphic.

Figure 4.4 is a relation graph including 5 nodes. Here nodes do not represent particular meanings, but one node is different from the others. The accessible relation in abstract way is R_1.

Fig. 4.4 Relation graph with 5 nodes

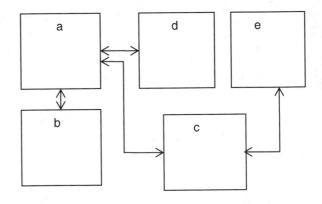

$$R_1 = \begin{bmatrix} from/to & a & b & c & d & e \\ a & 0 & 1 & 1 & 1 & 0 \\ b & 1 & 0 & 0 & 0 & 0 \\ c & 1 & 0 & 0 & 0 & 1 \\ d & 1 & 0 & 0 & 0 & 0 \\ e & 0 & 0 & 1 & 0 & 0 \end{bmatrix}$$

We remark that the matrix represents nodes that receive edges or send edges. So we begin with one node in the column that is the initial node of the edge and another node in the row receiving the edge. In fact we have the edges $e_{i,j}$ or $a_i \rightarrow a_j$, where a_i is the starting node and a_j is the final node. We also remark that any node has two different functions, one is the final node and the other is the start node. So in accord with the table, any node can be represented in Fig. 4.5.

So the original graph of the data base in Fig. 4.4 can be drawn in this abstract way (Fig. 4.6).

Example 4.1 Given the relation $R = \begin{bmatrix} R & a & b & c \\ a & 0 & 1 & 1 \\ b & 1 & 0 & 0 \\ c & 1 & 0 & 0 \end{bmatrix}$

Figure 4.7 shows the entities that send information to other entities.

Figure 4.8 shows the entities that receive information from other entities.

So Relation R is split into two parts (Fig. 4.9).

Now we present the commutative graphs that include the relations R and its transformed relation by permutation or transformation of "out" and "in" of the entities in Fig. 4.10.

Where

$$A = \begin{bmatrix} 0 & 1 & 0 \\ 0 & 0 & 1 \\ 1 & 0 & 0 \end{bmatrix}, \quad B = \begin{bmatrix} 0 & 1 & 0 \\ 1 & 0 & 0 \\ 0 & 0 & 1 \end{bmatrix}$$

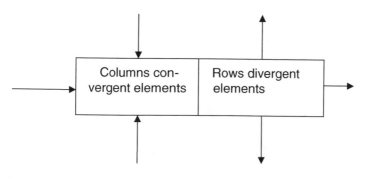

| Columns con-vergent elements | Rows divergent elements |

Fig. 4.5 Convergent and divergent part of the same entity

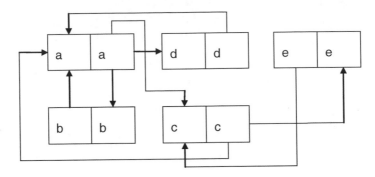

Fig. 4.6 Coherent graph that represents the rows and columns in the R_1

Fig. 4.7 Relation that represents to send information to other entities

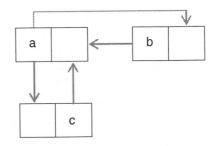

Fig. 4.8 Relation that represents to receive information from other entities

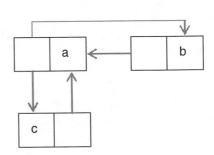

The previous diagram can be redrawn in Fig. 4.11.

And the relation T is given by the inconsistent graph Fig. 4.12.

Where the same entity has two different names. Now we change the relationship in a way to transform an inconsistent or conflicting graph into a consistent graph. So we have the relation T.

$$T = \begin{bmatrix} T & b & a & c \\ b & 0 & 1 & 1 \\ c & 1 & 0 & 0 \\ a & 1 & 0 & 0 \end{bmatrix}$$

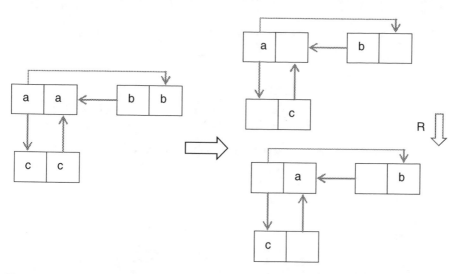

Fig. 4.9 Separation of a graph into sources and sinks

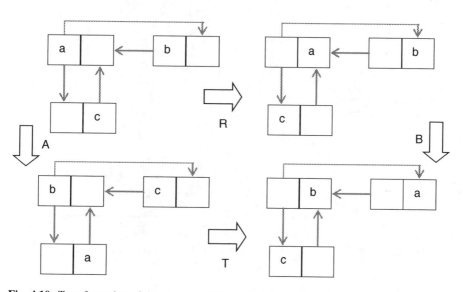

Fig. 4.10 Transformation of the sources entities and sinks entities

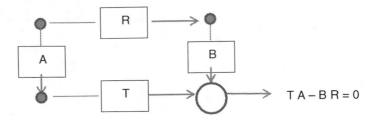

Fig. 4.11 Commutative graph for the transformation of the sources and the transformation of the sinks

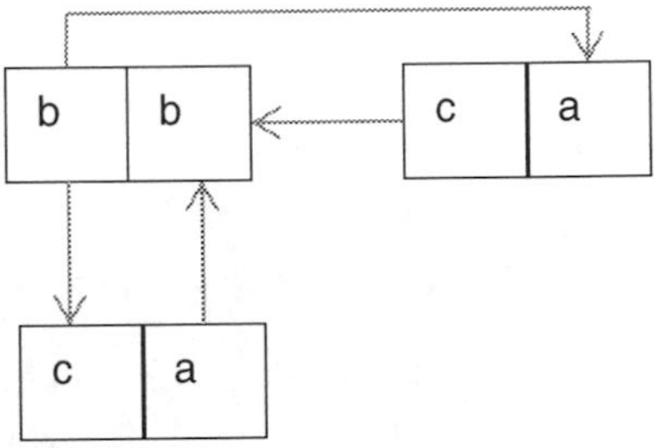

Fig. 4.12 Incoherent graph where the sources and the sinks for the same entity has two different names. The entity has two different names one in conflict with the other. The links are always the same but the entities are incoherent

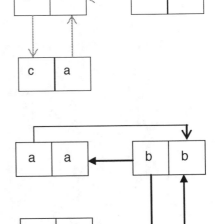

Fig. 4.13 Change of the links to generate a coherent graph for which any entity source and sink has the same name

When we order the rows and columns we have the consistent relation T (4.1).

$$
\begin{bmatrix}
T & b & a & c \\
b & 0 & 1 & 1 \\
c & 1 & 0 & 0 \\
a & 1 & 0 & 0
\end{bmatrix}
\rightarrow
\begin{bmatrix}
T & b & a & c \\
a & 1 & 0 & 0 \\
b & 0 & 1 & 1 \\
c & 1 & 0 & 0
\end{bmatrix}
\rightarrow
\begin{bmatrix}
T & a & b & c \\
a & 0 & 1 & 0 \\
b & 1 & 0 & 1 \\
c & 0 & 1 & 0
\end{bmatrix}
\tag{4.1}
$$

Figure 4.13 shows the coherent graph after the two different permutations.

From conflicting situation we can return to coherent state by a compensation process that changes the position of the relationship. Table 4.1 is a coherent table with a new morpho. So the change of the rows and columns is compensatory operation that changes an incoherent or conflicting situation with the same graph or

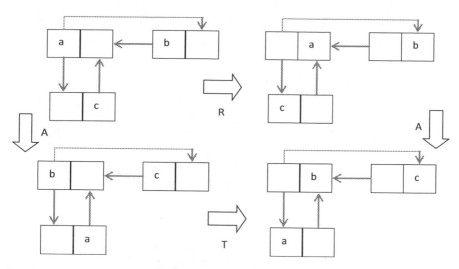

Fig. 4.14 Change of sources and sinks

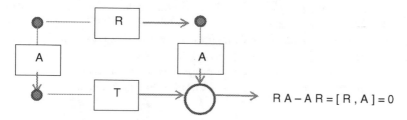

Fig. 4.15 Commutative graph for Fig. 4.14

Fig. 4.16 Coherent graph
with sinks and sources for any
entity the same name

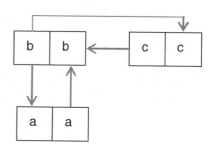

morpho into a new graph. The incoherence means that the permutation of the "out"
element with fixed "in" element in the same morpho or graph generate defects or
errors or conflicts between database and permutation. Then we have the commutative graph Fig. 4.14.

Figure 4.14 can be represented in this simple way (Fig. 4.15).

The relation T is shown in Fig. 4.16.

The difference between R and T is that with the same graph the entities are coherent but permuted. So R and T are similar or isomorphic. For the table relation we have

$$T = \begin{bmatrix} T & b & c & a \\ b & 0 & 1 & 1 \\ c & 1 & 0 & 0 \\ a & 1 & 0 & 0 \end{bmatrix}$$

That is consistent because any entity has only one name for the two parts. One goes to other entities, the other arrives from other entities. So we have (4.2).

$$\begin{bmatrix} T & b & c & a \\ b & 0 & 1 & 1 \\ c & 1 & 0 & 0 \\ a & 1 & 0 & 0 \end{bmatrix} \rightarrow \begin{bmatrix} T & b & c & a \\ a & 1 & 0 & 0 \\ b & 0 & 1 & 1 \\ c & 1 & 0 & 0 \end{bmatrix} \rightarrow \begin{bmatrix} T & a & b & c \\ a & 0 & 1 & 0 \\ b & 1 & 0 & 1 \\ c & 0 & 1 & 0 \end{bmatrix} \quad (4.2)$$

In a graph way we have the coherent graph after the two different permutations. The relation T is not equal to relation R but is equivalent. We have coherence that the graph change of the database does not lead to the change of the internal morpho. This means that the new database with a new organization of accessible relations is different because it has a different meaning, however, the new meaning or accessibility has the same morpho or internal properties. In another word, any reasoning or path of questions and answers is not equal but has the same structure.

4.4   From Inconsistent to Consistent Data by Map Reduction in Big Data

Any cluster in parallel with the others is separate in clusters with the same attributes as color or shuffle. The last part took cluster with the same color and built big clusters with the same attribute or color. This part is denoted reduction. Now the incoherent set of data is transformed in a set of coherent data by reduction (Fig. 4.17).

Table of colors as attributes and objects is the cluster of coherent data.

$$\begin{bmatrix} values & Red & Yellow & Blue \\ data_1 & a_{11} & a_{12} & a_{13} \\ \dots & \dots & \dots & \dots \\ data_n & a_{1n} & a_{2n} & a_{3n} \end{bmatrix}$$

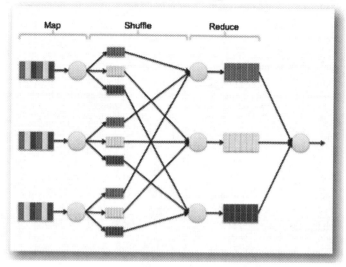

Fig. 4.17 Big data with its attribute in different colors is separate in cluster or map (Color figure online)

Fig. 4.18 With map reduction we can fix the big data main reference where we can create any type of geometry and trans formation

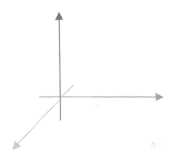

The map reduction creates the fundamental reference for the big data color space image or space of attribute as we can see in Figs. 4.18 and 4.19.

4.5 Simple Electrical Circuit as Database Graph Structure

The electrical circuit graph is made by two cycles joining one with the other by the resistor R_2. So the graph of the electrical circuit can, in a schematic way, is given by the structure in Figs. 4.8, 4.20 and 4.21.

The relation between nodes of the electrical circuit (topology) is given by the matrix 4.3.

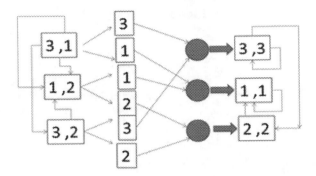

Fig. 4.19 Given a graph with sinks and sources that is incoherent. The same entity or point has two different names for the sink and source state. Now we know that any point or entity is a unity that has only one name. This is the incoherence condition as we have in the big data structure. Now with the map reduction we can transform the two dimensional incoherent graphs into a coherent graph for which any point is a cluster or sources and sinks with the same name as attribute. So we have a morpho transformation

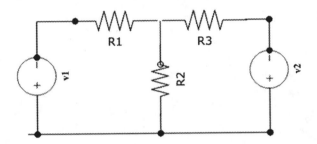

Fig. 4.20 Electrical network with electrical generator (active part), flow of currents and resistors

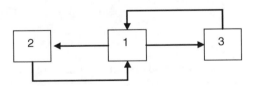

Fig. 4.21 Graphic scheme of the electrical circuit where we have two cycles. The node 2 is the first generator and the resistor R_1. The node 3 is the generator with resistor R_3, the node 1 is the resistor R_2

$$\begin{bmatrix} & node_1 & node_2 & node_3 \\ node_1 & 0 & 1 & 0 \\ node_2 & 1 & 0 & 1 \\ node_3 & 0 & 1 & 0 \end{bmatrix} \tag{4.3}$$

Now with the source, sink space we define the incident matrix or relation between nodes in this way (4.4).

$$
\begin{bmatrix} 1 & 0 & 0 \\ 0 & 1 & 0 \\ 0 & 0 & 1 \end{bmatrix} e_1 + \begin{bmatrix} 0 & 1 & 0 \\ 1 & 0 & 1 \\ 0 & 1 & 0 \end{bmatrix} e_2 = \begin{bmatrix} e_1 & e_2 & 0 \\ e_2 & e_1 & e_2 \\ 0 & e_2 & e_1 \end{bmatrix} \tag{4.4}
$$

The second row of the matrix indicates that in the electrical circuit graph we have one source and two sinks. Because we want to eliminate the bifurcation element we split the second row into two rows with one source and only one sink. So we have the formal transformation of the matrix in this way (4.5).

$$
\begin{bmatrix} e_1 & e_2 & 0 \\ e_2 & e_1 & e_2 \\ 0 & e_2 & e_1 \end{bmatrix} \rightarrow \begin{bmatrix} e_1 & e_2 & 0 \\ e_2 & e_1 & 0 \\ 0 & e_1 & e_2 \\ 0 & e_2 & e_1 \end{bmatrix} \tag{4.5}
$$

Now we compute the possible trajectories in the network by 4.6.

$$
\begin{bmatrix} e_1(1) & e_2(2) & 0 \\ e_2(4) & e_1(3) & 0 \\ 0 & e_1 & e_2 \\ 0 & e_2 & e_1 \end{bmatrix}, \begin{bmatrix} e_1 & e_2 & 0 \\ e_2 & e_1 & 0 \\ 0 & e_1(1) & e_2(2) \\ 0 & e_2(4) & e_1(3) \end{bmatrix} \tag{4.6}
$$

So the two trajectories are cycles. For any cycle includes a generator, which is an external element of the circuit that takes energy from external source and introduces this energy into the electrical circuit, the cycle is not closed. So we open any cycle in a single path with initial element or pure source, one transit element with one source and one sink, and one pure sink. So we have the path (Figure 4.22).

Where the first row is the connection between the pure source and the sink of the transit node. The second row is the connection between the source of the transit and the pure sink of the path. In Fig. 4.9 we have the transformation of a cycle into a path. In a matrix form we have the row reduction (4.7).

$$
\begin{bmatrix} e_1 & e_2 & 0 \\ e_2 & e_1 & 0 \\ 0 & e_1 & e_2 \\ 0 & e_2 & e_1 \end{bmatrix} \rightarrow \begin{bmatrix} e_1 & e_2 & 0 \\ 0 & e_1 & e_2 \end{bmatrix} \tag{4.7}
$$

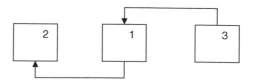

Fig. 4.22 The graph in Fig. 4.8 with two cycles is transformed in one path that moves from 3 to 1 and then to 2

In the original matrix, because we want to open the cycles we eliminate the second row and the fourth row. So the two cycles become two links one for any cycle. When we want to put in evidence the two links of the cycles and the flows in the rows we take the transport matrix of the reduction matrix so we have 4.8.

$$
\begin{bmatrix} e_1 & 0 \\ e_2 & e_1 \\ 0 & e_2 \end{bmatrix} \rightarrow
\begin{array}{c@{}c} & \begin{matrix} c_1 & c_2 \end{matrix} \\ \begin{bmatrix} 1 & 1 & 0 \\ 2 & 1 & 1 \\ 3 & 0 & 1 \end{bmatrix} \end{array}
\tag{4.8}
$$

where c_1 and c_2 are the links that substitutes the cycles. Now for the flows and voltages Kirchhoff laws we have 4.9.

$$
\begin{bmatrix} i_1 \\ i_1 + i_2 = i \\ i_2 \end{bmatrix} = \begin{bmatrix} 1 & 0 \\ 1 & 1 \\ 0 & 1 \end{bmatrix} \begin{bmatrix} i_1 \\ i_2 \end{bmatrix},
$$

$$
\begin{bmatrix} E_1 \\ E_2 \end{bmatrix} = \begin{bmatrix} 1 & 1 & 0 \\ 0 & 1 & 1 \end{bmatrix} \begin{bmatrix} v_1 \\ v \\ v_2 \end{bmatrix} = \begin{bmatrix} v_1 + v \\ v + v_2 \end{bmatrix}
\tag{4.9}
$$

We can see that the connection matrix is the same matrix generated by the source, sink space when we eliminate the cycle and substitute it for links and paths. Now in the path we separate the three entities in two parts, one for the sources and the other for the sink so we have Fig. 4.23.

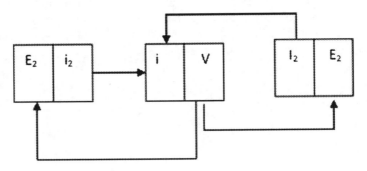

Fig. 4.23 For the voltages we have one path and for the current we have the reverse path

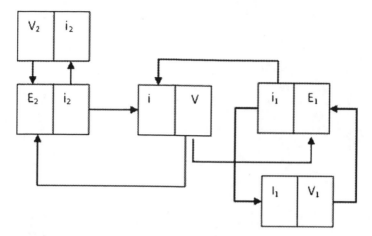

Fig. 4.24 Bond graph by sources and sinks any entity is a power system where the currents and voltages are connected as sinks and sources in agreement with Kirchhoff and path in the entity relation

Now for the same generator we have two resistors R_2, R_3 with the same current so we improve the previous graph with the two resistors in Fig. 4.24.

Chapter 5
Formal Description and References in Graph Theory

5.1 Formal Description of Relationships

Given the relation R

$$R = \begin{bmatrix} 0 & 1 & 1 \\ 1 & 0 & 0 \\ 1 & 0 & 0 \end{bmatrix}$$

We split the relation into two parts shown in Fig. 5.1.

The first part is the start part and the second part is the final part. Now the first part is the vector $\begin{bmatrix} a \\ b \\ c \end{bmatrix}$ the second part is the vector $\begin{bmatrix} (b,c) \\ a \\ a \end{bmatrix}$. Formally we can assume that the first vector is the values of coordinate e_1 and the second vector is the values of the coordinates e_2. So the graph can be represented symbolically as follows (5.1).

$$R = \begin{bmatrix} a \\ b \\ c \end{bmatrix} e_1 + \begin{bmatrix} (b,c) \\ a \\ a \end{bmatrix} e_2 \tag{5.1}$$

Or we can also write it as (5.2).

$$R = \left(\begin{bmatrix} 1 & 0 & 0 \\ 0 & 1 & 0 \\ 0 & 0 & 1 \end{bmatrix} e_1 + \begin{bmatrix} 0 & 1 & 1 \\ 1 & 0 & 0 \\ 1 & 0 & 0 \end{bmatrix} e_2 \right) \begin{bmatrix} a \\ b \\ c \end{bmatrix} = \begin{bmatrix} ae_1 + (b+c)e_2 \\ be_1 + ae_2 \\ ce_1 + ae_2 \end{bmatrix} \tag{5.2}$$

For the commutative diagram we have Fig. 5.2.

We have the formal description of the commutative diagram as (5.3).

© Springer International Publishing AG 2017
G. Resconi et al., *Introduction to Morphogenetic Computing*,
Studies in Computational Intelligence 703, DOI 10.1007/978-3-319-57615-2_5

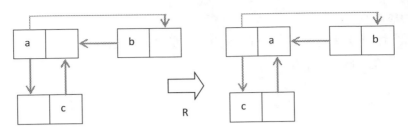

Fig. 5.1 Separation of the graph (relation) in sources and sinks

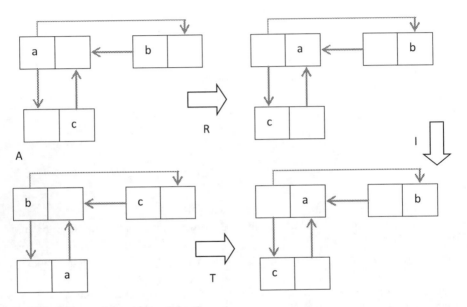

Fig. 5.2 Sources, sinks and transformations

$$\left(\begin{bmatrix} 1 & 0 & 0 \\ 0 & 1 & 0 \\ 0 & 0 & 1 \end{bmatrix} Ae_1 + \begin{bmatrix} 0 & 1 & 1 \\ 1 & 0 & 0 \\ 1 & 0 & 0 \end{bmatrix} e_2 \right) \begin{bmatrix} a \\ b \\ c \end{bmatrix} = \begin{bmatrix} Aae_1 + (b+c)e_2 \\ Abe_1 + ae_2 \\ Ace_1 + ae_2 \end{bmatrix}$$
$$= \begin{bmatrix} be_1 + (b+c)e_2 \\ ce_1 + ae_2 \\ ae_1 + ae_2 \end{bmatrix} \tag{5.3}$$

    The change of the initial value e_1 does not lead to the change of the final value which is comparable at the vector convergent graph Fig. 5.3.

    We can represent the previous initial value transformation also in this way (Fig. 5.4).

    And for the graph transformations we have (5.4) and (5.5).

Fig. 5.3 Change of the
source with the same sink

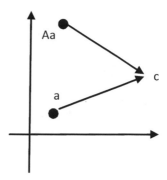

Fig. 5.4 The initial value
transformation

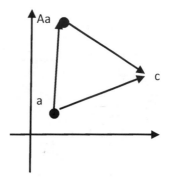

$$
Ae_1 \begin{bmatrix} a \\ b \\ c \end{bmatrix} = \begin{bmatrix} 0 & 1 & 0 \\ 0 & 0 & 1 \\ 1 & 0 & 0 \end{bmatrix} e_1 \begin{bmatrix} a \\ b \\ c \end{bmatrix} = \left(\begin{bmatrix} 1 & 0 & 0 \\ 0 & 1 & 0 \\ 0 & 0 & 1 \end{bmatrix} e_1 + \begin{bmatrix} 0 & 1 & 0 \\ 0 & 0 & 1 \\ 1 & 0 & 0 \end{bmatrix} e_2 \right) e_1 \begin{bmatrix} a \\ b \\ c \end{bmatrix}
$$

$$
= \left(\begin{bmatrix} 1 & 0 & 0 \\ 0 & 1 & 0 \\ 0 & 0 & 1 \end{bmatrix} e_1 + \begin{bmatrix} 0 & 1 & 0 \\ 0 & 0 & 1 \\ 1 & 0 & 0 \end{bmatrix} e_2 \right) e_1 \begin{bmatrix} a \\ b \\ c \end{bmatrix}
$$

(5.4)

$$
\left(\begin{bmatrix} 1 & 0 & 0 \\ 0 & 1 & 0 \\ 0 & 0 & 1 \end{bmatrix} Ae_1 + \begin{bmatrix} 0 & 1 & 1 \\ 1 & 0 & 0 \\ 1 & 0 & 0 \end{bmatrix} e_2 \right) \begin{bmatrix} a \\ b \\ c \end{bmatrix}
$$

(5.5)

$$
= \left(\left(\begin{bmatrix} 1 & 0 & 0 \\ 0 & 1 & 0 \\ 0 & 0 & 1 \end{bmatrix} e_1 + \begin{bmatrix} 0 & 1 & 0 \\ 0 & 0 & 1 \\ 1 & 0 & 0 \end{bmatrix} e_2 \right) e_1 + \begin{bmatrix} 0 & 1 & 1 \\ 1 & 0 & 0 \\ 1 & 0 & 0 \end{bmatrix} e_2 \right) \begin{bmatrix} a \\ b \\ c \end{bmatrix}
$$

And we have the commutative Fig. 5.5.
If we have the formal description as 5.6.

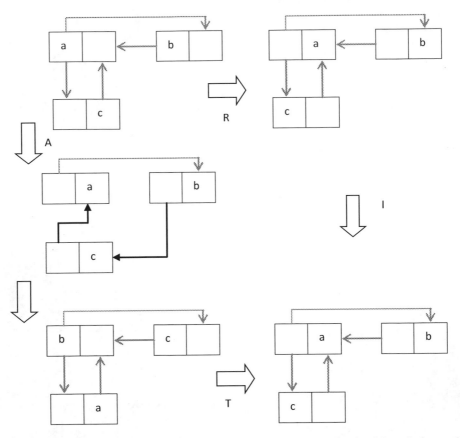

Fig. 5.5 Split of the transformation A first we transform the sources, after the sinks and after we have the final transformation A

$$\left(\begin{bmatrix} 1 & 0 & 0 \\ 0 & 1 & 0 \\ 0 & 0 & 1 \end{bmatrix} e_1 + \begin{bmatrix} 0 & 1 & 1 \\ 1 & 0 & 0 \\ 1 & 0 & 0 \end{bmatrix} Be_2\right) \begin{bmatrix} a \\ b \\ c \end{bmatrix} = \begin{bmatrix} ae_1 + B(b+c)e_2 \\ be_1 + Bae_2 \\ ce_1 + Bae_2 \end{bmatrix}$$
$$= \begin{bmatrix} ae_1 + (c+a)e_2 \\ be_1 + be_2 \\ ce_1 + be_2 \end{bmatrix} \tag{5.6}$$

We have it shown in Fig. 5.6.
And Fig. 5.7 shows the non-coherent graph.
Or in an explicit way we have the conflicting graph Fig. 5.8.

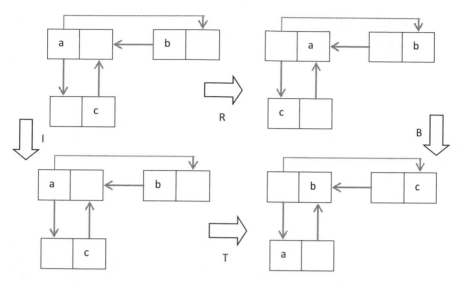

Fig. 5.6 Change of the sinks with the same sources

Fig. 5.7 The non-coherent graph given by the transformation of the sinks

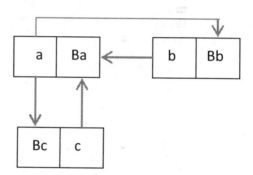

Fig. 5.8 The conflicting situation

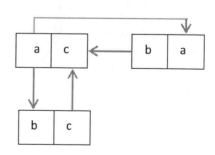

Fig. 5.9 Non-conflicting
situation

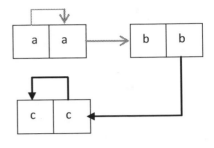

Fig. 5.10 With the same
sources, we change the sinks

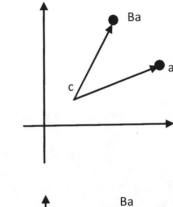

Fig. 5.11 With the same
source we change the sinks by
a vector sum

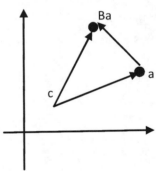

Now we change the relation in a way to have compensation and non-conflicting
graph Fig. 5.9.

With the analogy of the vector representation we have Fig. 5.10.

Or in a more complete way we have Fig. 5.11.

At the end we have 5.7

$$\left(\begin{bmatrix} 1 & 0 & 0 \\ 0 & 1 & 0 \\ 0 & 0 & 1 \end{bmatrix} Ae_1 + \begin{bmatrix} 0 & 1 & 1 \\ 1 & 0 & 0 \\ 1 & 0 & 0 \end{bmatrix} Ae_2 \right) \begin{bmatrix} a \\ b \\ c \end{bmatrix}$$

$$= \left(\begin{bmatrix} 1 & 0 & 0 \\ 0 & 1 & 0 \\ 0 & 0 & 1 \end{bmatrix} e_1 + \begin{bmatrix} 0 & 1 & 1 \\ 1 & 0 & 0 \\ 1 & 0 & 0 \end{bmatrix} e_2 \right) A \begin{bmatrix} a \\ b \\ c \end{bmatrix}$$

$$= \begin{bmatrix} Aae_1 + A(b+c)e_2 \\ Abe_1 + Aae_2 \\ Ace_1 + Aae_2 \end{bmatrix} = \begin{bmatrix} be_1 + (a+b)e_2 \\ ce_1 + ae_2 \\ ae_1 + ae_2 \end{bmatrix} \qquad (5.7)$$

So the relation $R = \left(\begin{bmatrix} 1 & 0 & 0 \\ 0 & 1 & 0 \\ 0 & 0 & 1 \end{bmatrix} e_1 + \begin{bmatrix} 0 & 1 & 1 \\ 1 & 0 & 0 \\ 1 & 0 & 0 \end{bmatrix} e_2 \right)$ is the same but we

change the name of the entities in this way $A \begin{bmatrix} a \\ b \\ c \end{bmatrix}$ and we have Fig. 5.12.

The graph has no conflicts. In the explicit way we have Fig. 5.13.

The relationships are the same and are coherent with the new graph when we only change the name of the entities. With coordinates we have Fig. 5.14.

Where the change moves from one vector (relationship) to another equivalent or parallel.

Fig. 5.12 Change in the same way the sources and the sinks

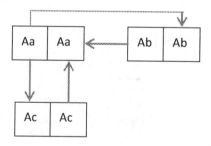

Fig. 5.13 When we change the sources and the sinks in the same way, the graph is always coherent. Source and sink of the same entity has the same name

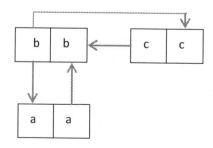

Fig. 5.14 Parallel
transformation when we
change the sources and the
sink in the same way

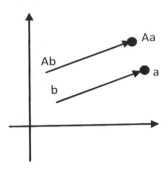

5.2   Topological Inconsistency

Given the path on the sphere Fig. 5.15.

Where the tangent vector is comparable with the relationship in the data base. After the cycle we can not return to the same relationship (tangent). This means that there is a conflict between the initial point and the final point. The mining of this conflict is that we use the same relation R that we use in three dimensional space where there is not constraint for the surface where we have constraint. So on the surface of the sphere we cannot have information of the curvature. This constraint as curvature is beyond our possibility and this creates the conflict. Now if we make the same cycle in the three dimensional space without constraint, the conflict will disappear. In conclusion, conflict appears because we use representations that do not take care of the constraints.

Fig. 5.15 The vector in the
cycle on the sphere cannot
return to the same vector so
for the point *A* the sink and
the source for the same point
or entity are different. This
creates an incoherent graph

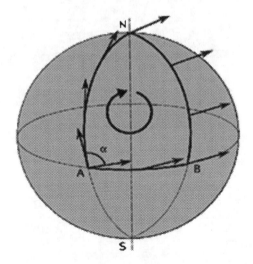

5.3   Inconsistency in Crystal Structure

Figure 5.16a, b is another non-coherent situation. In Fig. 5.16a the ideal network is mapped in crystal with curvature so we have incoherence in a loop similar to the sphere incoherence for vectors so we have the disclination defect in crystal structure. In Fig. 5.16b in a loop the same point is split into two parts, one for the source and the other for the sink. This defect is denoted dislocation because the same entity or point is dislocated in two different parts.

There is a sphere and a cylinder (Fig. 5.17).

The sphere cannot reduce to a cylinder because any vector that starts to a point or source after a loop becomes a sink vector whose direction is different from the direction of the source vector. In the cylinder the source and sink vectors after a loop are equal so at any point we have coherent loops.

When any entity is a point, the local coordinates show the three points where we can move to join one point with another point or one entity with another entity. Now we can see that locally there is no difference between the cylinder and the sphere. But globally one is not similar to the other. In fact at the polo Nord and Sud the two databases are completely different. We remark that at the same point Polo we have many different directions.

Let's take an example to explain the conflict situation. Given an elastic ring (Fig. 5.18), the ring can be seen as the joining together of infinite points.

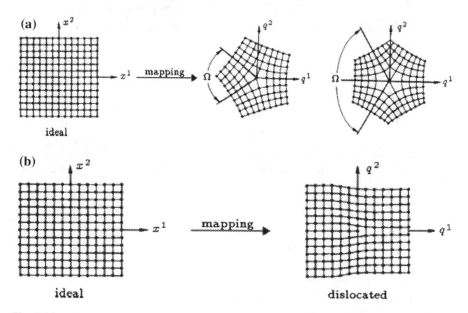

Fig. 5.16 a Ω is the angle between the vector source and the vector sink in the same point as in the sphere (dislocation). **b** In a loop the same point is split in two parts one for the source and the other for the sink

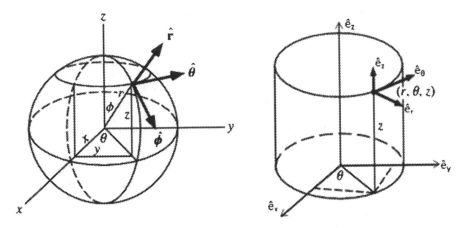

Fig. 5.17 The sphere and cylinder

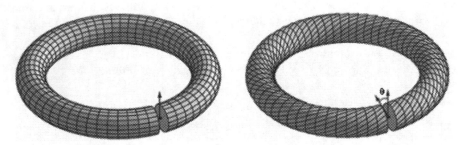

Fig. 5.18 Two rings are cut in one part. At the *left* the two cut surfaces are under the same force but at the *right* the two surface are under two different torque forces. So the source and the sink forces are coherent when they are under the same force, and are incoherent when they are under two different forces. So at the *left* for the same surface we have a stress situation that is the mechanical image of the inconsistency in logic

Originally the ring is a coherent structure and has no conflict. If we cut the ring at some joint, the connection of points is broken and two planes around the cutting point emerge. When we rotate the two planes with a certain angle at the same time, the ring is deformed and uncertainty is generated which decides what geometry topology the circle will be twisted. The uncertainty can be seen as dilemma or conflict. For small values of the twist, the ring will be deformed a little bit without out of two dimension space. For sufficiently high twist, the elastic ring will start writhing out of the plane which all composing points change into three-dimension space. As it is shown in Fig. 5.19.

To eliminate the conflict or uncertainty, we can apply two compensating ways. One way is to give the ring some force or pressure to make it restore to the original dimension space. The other way is to keep the ring in the multi-dimensional space and make it reach the stable state when deformed (Fig. 5.20).

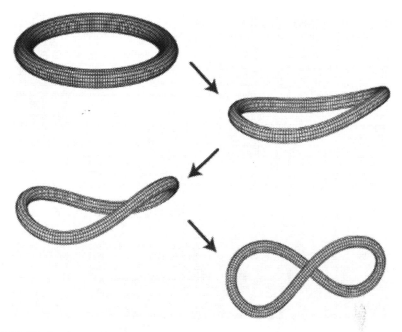

Fig. 5.19 The incoherent ring with internal stress is free to move to compensate the morpho or its form in a way to eliminate the internal stress or incoherence and give us the coherent ring with new forms. So the morpho or form changes in a way to give us coherent structure without internal force or stress. This is comparable with the database incoherence or conflict

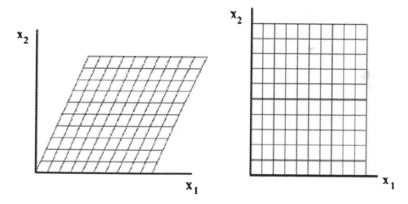

Fig. 5.20 In the transformation the source and sink for a point or entity are the same before and after the transformation. This is similar to the database transformation with the same operator for the source and the sink in any point or entity

In the global map or parallel transport, any close loop is transformed in a close loop. At any path before the transformation we have an equivalent path after the transformation. From coherent loop we move to another coherent loop. Any loop before and after the transformation is coherent.

Chapter 6
Logic of Conflicts and Active Set with Uncertainty and Incoherence

An active set is a unifying space being able to act as a "bridge" for transferring information, ideas and results between distinct types of uncertainties and different types of applications. An active set is a set of agents who independently deliver true or false values for a given proposition. An active set is not a simple vector of logic values for different propositions, the results are a vector but the set is not. The difference between an ordinary set and active set is that the ordinary set has passive elements with values of the attributes defined by an external agent. In the active set, any element is an agent that internally defines the value of a given attribute for a passive element. Agents in the active set with a special criterion give the logic value for the same attribute. So agents in many cases are in a logic conflict and this generates semantic uncertainty on the logic evaluation. Criteria and agents are the two variables by which we give different logic values to the same attribute or proposition. Active set is beyond the modal logic. In fact, given a proposition in modal logic we can evaluate the proposition only when we know the worlds where the proposition is located. When we evaluate one proposition in one world we cannot evaluate the same proposition in another world. Now in epistemic logic any world is an agent that knows the proposition is true or false. The active set is a set of agents as in the epistemic logic but is different from modal logic because all the agents (worlds) are not separate but are joined in the evaluation of the given proposition. In active set for one agent and one criterion we have one logic value but for many agents and criteria the evaluation is not single true or false but is a matrix of true and false. This matrix is not only a logic evaluation as in the modal logic but gives us the conflicting structure of the active set evaluation. Matrix agent is the vector subspace of the true or false agent multidimensional space. Operations among active sets include operations in the traditional sets, with fuzzy set and rough set as special cases. The agents multi dimensional space to evaluate active set include also the Hilbert multidimensional space where it is possible to simulate quantum logic gate. New logic operations are possible as fuzzy gate operations and

G. Resconi et al., *Introduction to Morphogenetic Computing*,
Studies in Computational Intelligence 703, DOI 10.1007/978-3-319-57615-2_6

more complex operations as conflicting solving, consensus operations, syntactic inconsistency, semantic inconsistency and knowledge integration. In the space of the agents evaluations morphotronic geometric operations are the new frontier to model new types of computers, new type of model for wireless communications as cognitive radio. In conclusion, active set opens the new possibility and new models for the logic.

6.1   Agents and Logic in the Epistemic Logic

Epistemic logic is the logic which formalizes knowledge of agents. Among many applications it is used in game theories and economic behaviour in databases and in verifying cryptographic protocols shared knowledge, common knowledge. Epistemic logic is also known as the logic of knowledge, it deals with modalities, which are not part of traditional logic and which modify the meaning of a proposition. For instance such a modality is the knowledge modality: "agent Alice knows that…", written K. Alice. There is one knowledge modality Ki for each agent i, so when there are n agents, there are n knowledge modalities. From the Ki's, one can build two new modalities, namely a modality Eg of shared knowledge, which modifies a proposition p into a proposition Eg(p) which means that "everyone in the group g knows p" and a modality Cg of common knowledge. Cg(p) would say "p is known to everybody in the group g" in a very strong sense since knowledge about p is known at every level of knowledge. Slightly more precisely, if g is the group of agents and p is a proposition, Eg(p) is the conjunction over the i 2 g of the Ki(p) and Cg(p) means something like "everybody knows p and everybody knows that everybody knows p and… and everybody knows that everybody knows that everybody knows…that everybody knows p…" This infinite conjunction is handled by making Cg(p) a fix point. A typical example of common knowledge is traffic regulation. When, as a car driver, you enter an intersection you know that the person on your left will let you go, moreover you know that she knows that you have the right to go and you are sure (you know) that she will not go because she knows that you know that she knows that you have the right to go etc. Actually you pass an intersection with a car on your left, because there is a common knowledge between you as a driver and the driver of the other car on the rule of priority. But those who travel have experienced the variability of the common knowledge. Take a stop sign. In Europe it means that the person which has a stop sign will let the other to pass the intersection. In some countries, the stop sign is just a decoration of intersections. In the USA, the common knowledge is different since there are intersections of two crossing roads with four stop signs and this has puzzled more than one European. One main goal of epistemic logic is to handle properly those concepts of knowledge of an agent, shared knowledge and common knowledge. So we have the Epistemic logic evaluation 6.1.

$$K_\alpha(p) = \begin{pmatrix} Agent & 1 & 2 & 3 \\ Logic \text{ value} & true & false & true \end{pmatrix} \tag{6.1}$$

where the proposition p is true for the agent 1, false for the agent 2 and true for the agent 3. Any agent in epistemic logic is completely separate from the others and any evaluation is given only when we know the agents as worlds. No conflict is possible because the agents are not considered together but one at the time as for the world in the modal logic.

6.2 Concepts and Definitions of Active Set

In the previous historical background we have vector evaluations but without any conflict because one evaluation is apart from the others. So p can be true in one situation and false in another but the two situations are not superposed so no conflict is possible. Only in quantum mechanics we can have superposition of different states where the proposition p can be both false and true. Now in quantum gate we use superposition and inconsistency only when we want to make a massive parallel computations. But when we want to measure the computation result, the superposition collapses and we always come back to a total separation of the states that is in agreement with the consistent classical logic by which we can make computation in the Boolean algebra. Now in a recently works on the agents appears the possibility to have inconsistent and conflict logic system where we can choose the consensus situation to come back to the classical and consistent true or false logic from inconsistency and also knowledge integration where we can know the logic value for complex propositions. Recently Cognitive radio system uses inconsistency to have a wireless efficient system. The aim of this chapter is to define a new type of set, that includes classical set theory, fuzzy set, set in evidence theory and rough set.

6.3 Properties and Definition of the Active Set

Any active set is a set of superpose agents, and any agent gives a value for the same proposition p. Active set appears similar to the Epistemic logic evaluation but the difference is that it is connected with the superposition of the world or agents whose judgment is not related to one agent but to the set of agents. We recognize active set elements in the vote process where all agents together give votes for the same person. In general the vote process is a conflicting vote because we have positive and negative votes for the same person. In epistemic logic this is impossible because we want to know where is the agent that gives a positive or negative judgment and this is possible without any conflict because we know the name of the agents. In active set, the set of agents is independent of the name that gives the judgment that must be only one for the set of agents. Now when all the agents obtain a consensus, they together give the same logic value so the conflict disappears and we have the classical non conflicting situation. The same is for knowledge integration where agents must be taken to integrate its actions to create the wanted knowledge integration. So now we begin with the formal description of the active set theory. Given three agents with all possible sets of logic values (true, false) one for any agents. So at any set of agent we have a power set of all possible evaluation for the proposition p. For example given three agents, the active set is a set of three agents with 8 sets of possible logic values for the same proposition p (as 6.2).

$$\Omega(p) = \begin{pmatrix} Agent & 1 & 2 & 3 \\ Logic\ value & true & true & true \end{pmatrix}, \begin{pmatrix} Agent & 1 & 2 & 3 \\ Logic\ value & true & true & false \end{pmatrix},$$
$$\begin{pmatrix} Agent & 1 & 2 & 3 \\ Logic\ value & true & false & true \end{pmatrix}, \begin{pmatrix} Agent & 1 & 2 & 3 \\ Logic\ value & false & true & true \end{pmatrix},$$
$$\begin{pmatrix} Agent & 1 & 2 & 3 \\ Logic\ value & false & false & true \end{pmatrix}, \begin{pmatrix} Agent & 1 & 2 & 3 \\ Logic\ value & false & true & false \end{pmatrix}$$
$$\begin{pmatrix} Agent & 1 & 2 & 3 \\ Logic\ value & true & false & false \end{pmatrix}, \begin{pmatrix} Agent & 1 & 2 & 3 \\ Logic\ value & false & false & false \end{pmatrix}$$

$$\tag{6.2}$$

In a more formal way we have 6.3.

$$SS(p) = \{A, \Omega(p) | A = set\ of\ agents, \Omega(p) = power\ set\ 2^A\ of\ the\ evaluations\ \}$$
$$\tag{6.3}$$

Given the proposition p, we denote as Criteria C one of the possible evaluation p in the set $\Omega(p)$. For example with three agents we have eight criteria to evaluate the proposition itself so we can write 6.4.

$$\Omega(p, C_1) = \begin{pmatrix} Agent & 1 & 2 & 3 \\ Logic\ value & true & true & true \end{pmatrix},$$

$$\Omega(p, C_2) = \begin{pmatrix} Agent & 1 & 2 & 3 \\ Logic\ value & true & true & false \end{pmatrix},$$

$$\Omega(p, C_3) = \begin{pmatrix} Agent & 1 & 2 & 3 \\ Logic\ value & true & false & true \end{pmatrix},$$

$$\Omega(p, C_4) = \begin{pmatrix} Agent & 1 & 2 & 3 \\ Logic\ value & false & true & true \end{pmatrix},$$

$$\Omega(p, C_5) = \begin{pmatrix} Agent & 1 & 2 & 3 \\ Logic\ value & false & false & true \end{pmatrix}, \qquad (6.4)$$

$$\Omega(p, C_6) = \begin{pmatrix} Agent & 1 & 2 & 3 \\ Logic\ value & false & true & false \end{pmatrix}$$

$$\Omega(p, C_7) = \begin{pmatrix} Agent & 1 & 2 & 3 \\ Logic\ value & true & false & false \end{pmatrix},$$

$$\Omega(p, C_8) = \begin{pmatrix} Agent & 1 & 2 & 3 \\ Logic\ value & false & false & false \end{pmatrix}$$

We remark that the set of Criteria is a mathematical lattice. For the previous example we have Fig. 6.1.

Operations

The agents set A is an ordinary set with normal intersection union and complementary operator. For the logic evaluation we have two different operations.

Fig. 6.1 Lattice of the uncertainty with different criteria

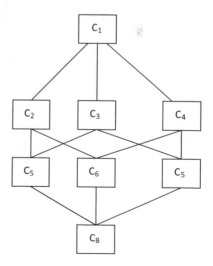

(1) Operation among criteria for the same proposition. Because we have the same proposition with two different criteria, we cannot compose the logic values that are heterogeneous. So we have the rule 6.5.

$$C_i \oplus C_j = \begin{pmatrix} Agent & 1 & 2 & \ldots & n \\ C_i & v_{1,1} & v_{1,2} & \ldots & v_{1,n} \\ C_j & v_{2,1} & v_{2,2} & \ldots & v_{2,n} \end{pmatrix} \tag{6.5}$$

So we increase the dimension of the space of the evaluation. For example, given ten agents and two criteria we have 6.6.

$$\Omega(p, C_i, C_j) = \begin{pmatrix} Agents & 1 & 2 & 3 & 4 & 5 \\ p, C_i & f & f & t & t & f \\ p, C_j & t & t & f & t & f \end{pmatrix} \tag{6.6}$$

In a graphic way we have Fig. 6.2.

(2) For two different propositions p and q we have the composition rule for the active set (as 6.7).

$$\Omega(p \wedge q, C) = \begin{pmatrix} Agents & 1 & 2 & \ldots & n \\ p, C & v_{1,p} & v_{2,p} & \ldots & v_{n,p} \end{pmatrix} \wedge \begin{pmatrix} Agents & 1 & 2 & \ldots & n \\ q, C & v_{1,q} & v_{2,q} & \ldots & v_{n,q} \end{pmatrix}$$
$$= \begin{pmatrix} Agents & 1 & 2 & \ldots & n \\ p, q, C & v_{1,p} \wedge v_{1,q} & v_{2,p} \wedge v_{2,q} & \ldots & v_{n,p} \wedge v_{n,q} \end{pmatrix} \tag{6.7}$$

$$\Omega(p \vee q, C) = \begin{pmatrix} Agents & 1 & 2 & \ldots & n \\ p, C & v_{1,p} & v_{2,p} & \ldots & v_{n,p} \end{pmatrix} \vee \begin{pmatrix} Agents & 1 & 2 & \ldots & n \\ q, C & v_{1,q} & v_{2,q} & \ldots & v_{n,q} \end{pmatrix}$$
$$= \begin{pmatrix} Agents & 1 & 2 & \ldots & n \\ p, q, C & v_{1,p} \vee v_{1,q} & v_{2,p} \vee v_{2,q} & \ldots & v_{n,p} \vee v_{n,q} \end{pmatrix} \tag{6.8}$$

Fig. 6.2 Two dimensional evaluation for two different criteria for five agents and the same proposition p

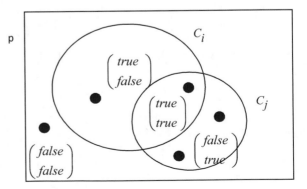

Example 6.1

$$\Omega(p) = \begin{pmatrix} Agents & 1 & 2 & 3 & 4 & 5 & 6 \\ values & t & t & t & f & f & f \end{pmatrix}, \quad \Omega(q) = \begin{pmatrix} Agents & 1 & 2 & 3 & 4 & 5 & 6 \\ values & t & t & t & t & f & f \end{pmatrix}$$

$$\Omega(p \vee q) = \begin{pmatrix} Agents & 1 & 2 & 3 & 4 & 5 & 6 \\ values & t & t & t & t & f & f \end{pmatrix},$$

$$\Omega(p \wedge q) = \begin{pmatrix} Agents & 1 & 2 & 3 & 4 & 5 & 6 \\ values & t & t & t & f & f & f \end{pmatrix}$$

The two logic operators are sensible to the order of the agents as a list for the negation operator we have 6.9.

$$\Omega(\neg p) = \begin{pmatrix} Agents & 1 & & 2 \\ value & \alpha_1(\neg v_1) + (1 - \alpha_1)(v_1) & & \alpha_2(\neg v_2) + (1 - \alpha_2)(v_2) \\ \dots & n & & \\ \dots & \alpha_n(\neg v_n) + (1 - \alpha_n)(v_n) & & \end{pmatrix} \tag{6.9}$$

Example 6.2

$$\Omega(p) = \begin{pmatrix} Agents & 1 & 2 & 3 & 4 & 5 & 6 \\ values & f & f & f & t & t & t \end{pmatrix}$$

For

if $\alpha = \begin{pmatrix} Agents & 1 & 2 & 3 & 4 & 5 & 6 \\ values & 1 & 1 & 1 & 1 & 1 & 1 \end{pmatrix}$ then $\Omega(\neg p) = \begin{pmatrix} Agents & 1 & 2 & 3 & 4 & 5 & 6 \\ values & t & t & t & f & f & f \end{pmatrix}$

if $\alpha = \begin{pmatrix} Agents & 1 & 2 & 3 & 4 & 5 & 6 \\ values & 1 & 1 & 1 & 0 & 1 & 1 \end{pmatrix}$ then $\Omega(\neg p) = \begin{pmatrix} Agents & 1 & 2 & 3 & 4 & 5 & 6 \\ values & t & t & t & t & f & f \end{pmatrix}$

if $\alpha = \begin{pmatrix} Agents & 1 & 2 & 3 & 4 & 5 & 6 \\ values & 1 & 1 & 0 & 1 & 1 & 1 \end{pmatrix}$ then $\Omega(\neg p) = \begin{pmatrix} Agents & 1 & 2 & 3 & 4 & 5 & 6 \\ values & t & t & f & f & f & f \end{pmatrix}$

When all the values of α are equal to one, all the agents change its value in the negation operation. When one α is zero for the true values one true value agent does not change and all the others change. So in the end the number of agents with true value in the negation operation is more than in the classical negation for any agent. On the contrary, if α is zero for one agent with false value, the number of the true value in the negation is less than the classical negation for any agent.

6.4 Aggregation Rule for Active Set

Given an active set, we associate to any active set evaluation a number by an aggregation function that can be linear or non linear. For the linear case the aggregation can be simple aggregation or can be weighted aggregation. For example for simple linear aggregation rule we have the aggregation rule 6.10.

$$for \ \Omega(p, C_1) = \begin{pmatrix} Agent & 1 & 2 & 3 \\ Logic \ value & true & true & true \end{pmatrix}$$

$$\mu(p, C_1) = \frac{1}{3}|true\rangle + \frac{1}{3}|true\rangle + \frac{1}{3}|true\rangle = \frac{1}{3} + \frac{1}{3} + \frac{1}{3} = 1$$

$$for \ \Omega(p, C_2) = \begin{pmatrix} Agent & 1 & 2 & 3 \\ Logic \ value & true & true & false \end{pmatrix}$$

$$\mu(p, C_2) = \frac{1}{3}|true\rangle + \frac{1}{3}|true\rangle + \frac{1}{3}|false\rangle = \frac{1}{3} + \frac{1}{3} + \frac{1}{3}0 = \frac{2}{3}$$

(6.10)

Where Q is the linear superposition of the logic value for the active set.

6.5 Fuzzy Set by Active Set

The probability calculus does not incorporate explicitly the concepts of irrationality or agent's state of logic conflict. It misses structural information at the level of individual objects, but preserves global information at the level of a set of objects. Given a dice the probability theory studies frequencies of the different faces $E = \{e\}$ as independent (elementary) events. This set of elementary events E has no structure. It is only required that elements of E be mutually exclusive and complete, and there is no other possible alternative. The order of its elements is irrelevant to probabilities of each element of E. No irrationality or conflict is allowed in this definition relative to mutual exclusion. The classical probability calculus does not provide a mechanism for modelling uncertainty when agents communicate (collaborates or conflict). Below we present the important properties of sets of conflicting agents at one dimension Let $\Omega(x)$ the active set for the proposition x and $|\Omega(x)|$ be the numbers of agents for which proposition x is true we have
Given two propositions a and b when

$$If \ |\Omega(a)| < |\Omega(b)| \ then \ p = a \ and \ q = b$$
$$If \ |\Omega(b)| < |\Omega(a)| \ then \ p = b \ and \ q = a$$

So we order the propositions from the proposition with less number of true value to the proposition with maximum of true values (Fig. 6.3)

Fig. 6.3 Fuzzy rules and
Active sets

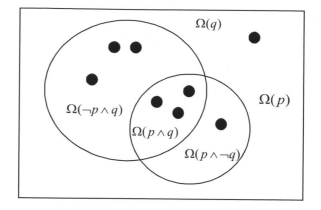

$$|\Omega(p)| = 4$$
$$|\Omega(q)| = 5$$
$$|\Omega(p)| < |\Omega(q)|$$
$$\max(|\Omega(p)|, |\Omega(q)|) = |\Omega(q)|$$
$$\min(|\Omega(p)|, |\Omega(q)|) = |\Omega(q)|$$

$$\Omega(p) = \begin{pmatrix} Agents & 1 & 2 & 3 & 4 & 5 & 6 & 7 & 8 \\ values & f & f & t & t & t & t & f & f \end{pmatrix}$$

$$\Omega(q) = \begin{pmatrix} Agents & 1 & 2 & 3 & 4 & 5 & 6 & 7 & 8 \\ values & f & t & t & f & t & t & f & t \end{pmatrix}$$

$$|\Omega(p)| = 4, \quad |\Omega(q)| = 5$$

We have

$$\Omega(p \wedge q) = \begin{pmatrix} Agents & 1 & 2 & 3 & 4 & 5 & 6 & 7 & 8 \\ values & f & f & t & f & t & t & f & f \end{pmatrix}$$

$$\Omega(p \vee q) = \begin{pmatrix} Agents & 1 & 2 & 3 & 4 & 5 & 6 & 7 & 8 \\ values & f & t & t & t & t & f & t, \end{pmatrix}$$

$$|\Omega(p \wedge q) = 3|, \quad |\Omega(p \vee q) = 6|$$

Now we know that

$$q \vee (p \wedge \neg q) = (q \vee p) \wedge (q \vee \neg q) = q \vee p$$
$$p \wedge \neg(p \wedge \neg q) = p \wedge (\neg p \vee q) = (p \wedge \neg p) \vee p \wedge q = p \wedge q$$

But because when q is false and p is true we adjoin one logic value true at q to obtain p or q. So when we repeat this process many times for any agent we have that at the number of true values for q we must adjoin other true values for which q is false but p is true. In conclusion we have

$|\Omega(p \vee q)| = |\Omega(q)| + |\Omega(\neg q \wedge p)| = \max(|\Omega(q)|, |\Omega(p)| + |\Omega(\neg q \wedge p)|$ For any operation we have that when q is false and p is true we eliminate one element for which p is true. In conclusion when we repeat this for many times we have

$$|\Omega(p \wedge q)| = |\Omega(p)| - |\Omega(\neg q \wedge p)| = \min(|\Omega(q)|, |\Omega(p)| + |\Omega(\neg q \wedge p)|$$

In one word, in the active set we can find the Zadeh rule again when p and not q is always false.

Zadeh rule

$$|\Omega(p \wedge q)| = \min(|\Omega(q)|, |\Omega(p)|$$
$$|\Omega(p \vee q)| = \max(|\Omega(q)|, |\Omega(p)|$$

So when the agents for which p is true are also the agents for which q is true. In a graphic way we have Fig. 6.4.

We can also remark that the minimum rule is the maximum possible value for AND and the maximum rule is the minimum possible value for OR. We can see that for the previous example we have

$$for \;\; |\Omega(p \wedge \neg q)| = 1$$
$$|\Omega(p \wedge q)| = \min(|\Omega(p)|, |\Omega(q)|) - |\Omega(p \wedge \neg q)| = 4 - 1 = 3$$
$$|\Omega(p \vee q)| = \max(|\Omega(p)|, |\Omega(q)|) + |\Omega(p \wedge \neg q)| = 5 + 1 = 6$$

For the negation we have the Zadeh rule

$$|\Omega(\neg p)| = n - |\Omega(p)|$$

When we divide agents with the number n, we have the traditional rule

$$\mu(\neg p) = \frac{|\Omega(\neg p)|}{n} = 1 - \frac{|\Omega(p)|}{n} = 1 - \mu(p)$$

Fig. 6.4 Zadeh fuzzy rules and active sets

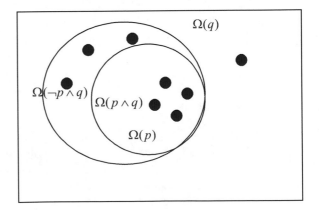

Fig. 6.5 Change of the Sugeno negation value for the change of lambda parameter

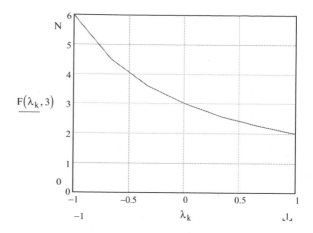

In this situation all the agents in the negation change all the logic values in a synchronic way. But when we have the Sugeno rule

$$|\Omega(\neg p)| = \mu(\neg p)n = \frac{1 - \mu(p)}{1 + \lambda\mu(p)}n = \frac{1 - \frac{|\Omega(p)|}{n}}{1 + \lambda\frac{|\Omega(p)|}{n}}n = n\frac{n - |\Omega(p)|}{n + \lambda|\Omega(p)|}$$

where $\lambda = [-1, \infty]$ when we change the lambda parameters for $n = 6$ and $\Omega(p) = 3$ we have the negation value (Fig. 6.5).

When $\lambda = 0$ all the agents change their logic values. So before we have three true values and three false values for the negation we have the same values again but are reversed. For

$$if \quad \lambda = 0, \quad |\Omega(\neg p)| = n - |\Omega(p)| = 6 - 3 = 3$$
$$if \quad \lambda < 0, \quad |\Omega(\neg p)| > n - |\Omega(p)|$$
$$if \quad \lambda > 0, \quad |\Omega(\neg p)| < n - |\Omega(p)|$$

When λ is negative, agents with true values do not change, when λ is positive, agents with false values do not change. In conclusion, t-norm and t-conorm and fuzzy negation can be simulates inside the active set.

6.6 Theory of Inconsistent Graph and Active Set

Given the inconsistent graph Fig. 6.6
We have the active set definition.

Fig. 6.6 The nodes are
inconsistent

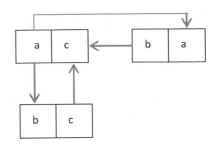

$$value(a_L) = \begin{bmatrix} agents & entity_1 & entity_2 & entity_3 \\ value & T & F & F \end{bmatrix}$$

$$value(b_L) = \begin{bmatrix} agents & entity_1 & entity_2 & entity_3 \\ value & F & T & F \end{bmatrix}$$

$$value(c_L) = \begin{bmatrix} agents & entity_1 & entity_2 & entity_3 \\ value & F & F & T \end{bmatrix}$$

$$value(a_R) = \begin{bmatrix} agents & entity_1 & entity_2 & entity_3 \\ value & F & T & F \end{bmatrix}$$

$$value(b_R) = \begin{bmatrix} agents & entity_1 & entity_2 & entity_3 \\ value & F & F & T \end{bmatrix}$$

$$value(c_R) = \begin{bmatrix} agents & entity_1 & entity_2 & entity_3 \\ value & T & F & F \end{bmatrix}$$

When we compose the left active sets with the right active set by logic equivalence operation we have

$$\begin{bmatrix} agents & entity_1 & entity_2 & entity_3 \\ value & T & F & F \end{bmatrix} = \begin{bmatrix} agents & entity_1 & entity_2 & entity_3 \\ value & F & T & F \end{bmatrix}$$
$$\rightarrow \begin{bmatrix} agents & entity_1 & entity_2 & entity_3 \\ value & F & F & T \end{bmatrix}$$

$$\begin{bmatrix} agents & entity_1 & entity_2 & entity_3 \\ value & F & T & F \end{bmatrix} = \begin{bmatrix} agents & entity_1 & entity_2 & entity_3 \\ value & F & F & T \end{bmatrix}$$
$$\rightarrow \begin{bmatrix} agents & entity_1 & entity_2 & entity_3 \\ value & T & F & F \end{bmatrix}$$

$$\begin{bmatrix} agents & entity_1 & entity_2 & entity_3 \\ value & F & F & T \end{bmatrix} = \begin{bmatrix} agents & entity_1 & entity_2 & entity_3 \\ value & T & F & F \end{bmatrix}$$
$$\rightarrow \begin{bmatrix} agents & entity_1 & entity_2 & entity_3 \\ value & F & T & F \end{bmatrix}$$

For consistent graph Fig. 6.7.

Fig. 6.7 The nodes are consistent

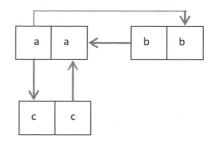

We have

$$\begin{bmatrix} agents & entity_1 & entity_2 & entity_3 \\ value & T & F & F \end{bmatrix} = \begin{bmatrix} agents & entity_1 & entity_2 & entity_3 \\ value & T & F & F \end{bmatrix}$$

$$\rightarrow \begin{bmatrix} agents & entity_1 & entity_2 & entity_3 \\ value & T & T & T \end{bmatrix}$$

$$\begin{bmatrix} agents & entity_1 & entity_2 & entity_3 \\ value & F & T & F \end{bmatrix} = \begin{bmatrix} agents & entity_1 & entity_2 & entity_3 \\ value & F & T & F \end{bmatrix}$$

$$\rightarrow \begin{bmatrix} agents & entity_1 & entity_2 & entity_3 \\ value & T & T & T \end{bmatrix}$$

$$\begin{bmatrix} agents & entity_1 & entity_2 & entity_3 \\ value & F & F & T \end{bmatrix} = \begin{bmatrix} agents & entity_1 & entity_2 & entity_3 \\ value & F & F & T \end{bmatrix}$$

$$\rightarrow \begin{bmatrix} agents & entity_1 & entity_2 & entity_3 \\ value & T & T & T \end{bmatrix}$$

For consistent graph the logic equivalence for active sets gives the value which is always true. So we have no conflicts.

Chapter 7
Cycles, Sinks, Sources and Links Products

In Chap. 1, we have given the representation of relation with sinks and sources in database. In this chapter, we use a new instrument to give a method for modelling a graph with cycles, sinks and sources by the external product.

Depending on the relationship, every entity in database can be split into two parts, the first part is the source, and the second one is the sink. How to represent the two parts in mathematics is a big problem. In Chap. 1, we gave the representation of relations as 7.1.

$$
\left(\begin{bmatrix} 1 & 0 & 0 & 0 & 0 \\ 0 & 1 & 0 & 0 & 0 \\ 0 & 0 & 1 & 0 & 0 \\ 0 & 0 & 0 & 1 & 0 \\ 0 & 0 & 0 & 0 & 1 \end{bmatrix} e_1 + \begin{bmatrix} 0 & 1 & 1 & 1 & 0 \\ 1 & 0 & 0 & 0 & 0 \\ 1 & 0 & 0 & 0 & 1 \\ 1 & 0 & 0 & 0 & 0 \\ 0 & 0 & 1 & 0 & 0 \end{bmatrix} e_2 \right) \begin{bmatrix} class \\ classroom \\ enrollment \\ teacher \\ student \end{bmatrix}
$$
$$
= \begin{bmatrix} (class)e_1 + (classroom + enrollment + teacher)e_2 \\ (classroom)e_1 + (class)e_2 \\ (enrollment)e_1 + (class + student)e_2 \\ (teacher)e_1 + (class)e_2 \\ (student)e_1 + (enrollment)e_2 \end{bmatrix} \tag{7.1}
$$

A more simple example is 7.2

$$
\left(\begin{bmatrix} 1 & 0 \\ 0 & 1 \end{bmatrix} e_1 + \begin{bmatrix} 1 & 1 \\ 1 & 1 \end{bmatrix} e_2 \right) \begin{bmatrix} a \\ b \end{bmatrix} = \begin{bmatrix} ae_1 + ae_2 + be_2 \\ be_1 + ae_2 + be_2 \end{bmatrix} \tag{7.2}
$$

The network can be represented in Fig. 7.1.

© Springer International Publishing AG 2017
G. Resconi et al., *Introduction to Morphogenetic Computing*,
Studies in Computational Intelligence 703, DOI 10.1007/978-3-319-57615-2_7

Fig. 7.1 Relations with each
node having two parts

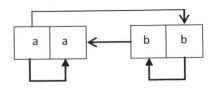

We have four links given by the vector of the links

$$
\begin{bmatrix} ae_1 + ae_2 \\ be_1 + be_2 \\ ae_1 + be_2 \\ be_1 + ae_2 \end{bmatrix} = \begin{bmatrix} f_1 \\ f_2 \\ f_3 \\ f_4 \end{bmatrix}
$$

We remark that $(e_1 + e_2)a$ is a self loop because the entity a includes a sink and a source together, the same for $(e_1 + e_2)b$. The tensor product of the vector is

$$
\begin{bmatrix} ae_1 + ae_2 \\ be_1 + be_2 \\ ae_1 + be_2 \\ be_1 + ae_2 \end{bmatrix} \begin{bmatrix} ae_1 + ae_2 \\ be_1 + be_2 \\ ae_1 + be_2 \\ be_1 + ae_2 \end{bmatrix}^T = \begin{bmatrix} (ae_1 + ae_2)(ae_1 + ae_2) & (ae_1 + ae_2)(be_1 + be_2) \\ (be_1 + be_2)(ae_1 + ae_2) & (be_1 + be_2)(be_1 + be_2) \\ (ae_1 + be_2)(ae_1 + ae_2) & (ae_1 + be_2)(be_1 + be_2) \\ (be_1 + ae_2)(ae_1 + ae_2) & (be_1 + be_2)(be_1 + ae_2) \end{bmatrix}
$$

$$
\begin{bmatrix} (ae_1 + ae_2)(ae_1 + be_2) & (ae_1 + ae_2)(be_1 + ae_2) \\ (be_1 + be_2)(ae_1 + be_2) & (be_1 + be_2)(be_1 + ae_2) \\ (ae_1 + be_2)(ae_1 + be_2) & (ae_1 + be_2)(be_1 + ae_2) \\ (be_1 + ae_2)(ae_1 + be_2) & (be_1 + ae_2)(be_1 + ae_2) \end{bmatrix}
$$

For the value $(ae_1 + ae_2)(be_1 + be_2)$ we have the connected conflicting graph Fig. 7.2.

But we know that the two links are not connected in fact. With compensation we have the consistent graph Fig. 7.3.

Fig. 7.2 Conflicting graph

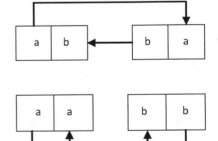

Fig. 7.3 Compensate graph

For the values

$$(ae_1 + be_2)(be_1 + ae_2) \rightarrow \begin{bmatrix} abe_1^2 \\ a^2e_1e_2 \\ b^2e_2e_1 \\ bae_2^2 \end{bmatrix} = \begin{bmatrix} sources \\ extreme \\ connection \\ \sin ks \end{bmatrix}$$

we have the consistent connected graph Fig. 7.4.

Now because the initial value ae_1 and the final value ae_2 have the same name, the previous value is a loop. When we build the graph of the consistent links we get the evolution connection table R.

$$\begin{bmatrix} R & f_1 & f_2 & f_3 & f_4 \\ f_1 & 1 & 0 & 1 & 0 \\ f_2 & 0 & 1 & 0 & 1 \\ f_3 & 1 & 0 & 0 & 1 \\ f_4 & 0 & 1 & 1 & 0 \end{bmatrix}$$

For this we have the second order graph Fig. 7.5 generated by two links connection.

Fig. 7.4 Consistent graph

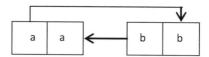

Fig. 7.5 Second order graph

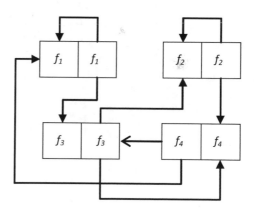

7.1   Study of Sink Property

For the relation shown in Fig. 7.6,
    We have the table

$$
\begin{bmatrix}
Data & (class)e_2 & (classroom)e_2 & (enrollment)e_2 & (teacher)e_2 & (student)e_2 \\
(class)e_1 & 0 & 1 & 1 & 1 & 0 \\
(classroom)e_1 & 0 & 0 & 0 & 0 & 0 \\
(enrollment)e_1 & 1 & 0 & 0 & 0 & 1 \\
(teacher)e_1 & 1 & 0 & 0 & 0 & 0 \\
(student)e_1 & 0 & 0 & 0 & 1 & 0
\end{bmatrix}
$$

There is no link starting from classroom to other entities, so classroom is a sink. We recognize the sink because all the row values of the sink are equal to zero. So we have

$$
\left(
\begin{bmatrix}
1 & 0 & 0 & 0 & 0 \\
0 & 1 & 0 & 0 & 0 \\
0 & 0 & 1 & 0 & 0 \\
0 & 0 & 0 & 1 & 0 \\
0 & 0 & 0 & 0 & 1
\end{bmatrix} e_1 +
\begin{bmatrix}
0 & 1 & 1 & 1 & 0 \\
0 & 0 & 0 & 0 & 0 \\
1 & 0 & 0 & 0 & 1 \\
1 & 0 & 0 & 0 & 0 \\
0 & 0 & 1 & 0 & 0
\end{bmatrix} e_2
\right)
\begin{bmatrix}
class \\
classroom \\
enrollment \\
theacher \\
student
\end{bmatrix}
$$

$$
=
\begin{bmatrix}
(class)e_1 + (classroom + enrollment + teacher)e_2 \\
(classroom)e_1 \\
(enrolment)e_1 + (class + student)e_2 \\
(teacher)e_1 + (class)e_2 \\
(student)e_1 + (enrollment)e_2
\end{bmatrix}
$$

Fig. 7.6 Relation in database

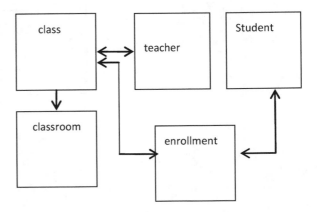

When we multiply for the sink we have

$$[(class)e_1 + (classroom)e_2][(classroom)e_1]$$

7.2   Study of Source Property

For the relation shown in Fig. 7.7, we have the table.

$$\begin{bmatrix} & class & classroom & enrollment & teacher & student \\ class & 0 & 1 & 1 & 0 & 0 \\ classroom & 1 & 0 & 0 & 0 & 0 \\ enrollment & 1 & 0 & 0 & 0 & 1 \\ teacher & 1 & 0 & 0 & 0 & 0 \\ student & 0 & 0 & 1 & 0 & 0 \end{bmatrix}$$

From teacher, we can go to class but no entity goes to teacher. All values of teacher column in the relation are equal to zero. So teacher is a source.
And

$$\left(\begin{bmatrix} 1 & 0 & 0 & 0 & 0 \\ 0 & 1 & 0 & 0 & 0 \\ 0 & 0 & 1 & 0 & 0 \\ 0 & 0 & 0 & 1 & 0 \\ 0 & 0 & 0 & 0 & 1 \end{bmatrix} e_1 + \begin{bmatrix} 0 & 1 & 1 & 0 & 0 \\ 1 & 0 & 0 & 0 & 0 \\ 1 & 0 & 0 & 0 & 1 \\ 1 & 0 & 0 & 0 & 0 \\ 0 & 0 & 1 & 0 & 0 \end{bmatrix} e_2 \right) \begin{bmatrix} class \\ classroom \\ enrollment \\ teacher \\ student \end{bmatrix}$$

$$= \begin{bmatrix} (class)e_1 + (classroom + enrollment)e_2 \\ (classroom)e_1 + (class)e_2 \\ (enrollment)e_1 + (class + student)e_2 \\ (teacher)e_1 + (class)e_2 \\ (student)e_1 + (enrollment)e_2 \end{bmatrix}$$

Fig. 7.7 Teacher is a source

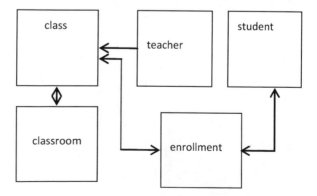

When we multiply for the source we have

$$[(teacher)e_1 + (class)e_2][(class)e_1 + (classroom + enrollment)e_2]$$

7.3 Both Sink and Source

The sink entity can be seen (accessible) from all the other entities but from the sink entity it is impossible to see the other entities. For the source entity we have dual property. From the source it is possible to see all the other entities but all the others cannot see the interior information of the source. Now in database it is possible to have sink and source entities together. Figure 7.8 is the example.

The from/to relation is

$$
\begin{bmatrix}
 & class & classroom & enrollment & teacher & student \\
class & 0 & 0 & 1 & 1 & 0 \\
classroom & 1 & 0 & 0 & 0 & 0 \\
enrollment & 1 & 0 & 0 & 0 & 1 \\
teacher & 0 & 0 & 0 & 0 & 0 \\
student & 0 & 0 & 1 & 0 & 0
\end{bmatrix}
$$

We can see that teacher is a sink and classroom is a source.

7.4 Cycle in the Database

When no sink or source is present in a graph or sub-graph, there must be cycles included. For example, in the following relation, there are one sink and one source.

Fig. 7.8 Sink and source together

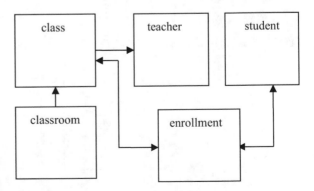

$$
\left(\begin{bmatrix} 1 & 0 & 0 & 0 & 0 \\ 0 & 1 & 0 & 0 & 0 \\ 0 & 0 & 1 & 0 & 0 \\ 0 & 0 & 0 & 1 & 0 \\ 0 & 0 & 0 & 0 & 1 \end{bmatrix} e_1 + \begin{bmatrix} 0 & 0 & 1 & 0 & 0 \\ 1 & 0 & 0 & 0 & 0 \\ 1 & 0 & 0 & 0 & 1 \\ 0 & 0 & 0 & 0 & 0 \\ 0 & 0 & 1 & 0 & 0 \end{bmatrix} e_2 \right) \begin{bmatrix} class \\ classroom \\ enrollment \\ teacher \\ student \end{bmatrix}
$$

$$
= \left(\begin{bmatrix} e_1 & 0 & e_2 & 0 & 0 \\ e_2 & e_1 & 0 & 0 & 0 \\ e_2 & 0 & e_1 & 0 & e_2 \\ 0 & 0 & 0 & e_1 & 0 \\ 0 & 0 & e_2 & 0 & e_1 \end{bmatrix} \begin{bmatrix} class \\ classroom \\ enrollment \\ teacher \\ student \end{bmatrix}\right)
$$

$$
= \begin{bmatrix} (class)e_1 + (enrollment)e_2 \\ (classroom)e_1 + (class)e_2 \\ (enrollment)e_1 + (class + student)e_2 \\ (teacher)e_1 \\ (student)e_1 + (enrollment)e_2 \end{bmatrix}
$$

Now in the sub-graph (class, enrollment), (student, enrollment), the table is

	class	classroom	enrollment	teacher	student
class	0	0	1	0	0
classroom	0	0	0	0	0
enrollment	1	0	0	0	0
teacher	0	0	0	0	0
student	0	0	0	0	0

	class	classroom	enrolment	teacher	student
class	0	0	1	0	0
classroom	0	0	0	0	0
enrolment	0	0	0	0	1
teacher	0	0	0	0	0
student	0	0	0	0	0

There is no sink and source. So we have the cycles.

$$
\begin{bmatrix} e_1(1) & 0 & e_2(2) & 0 & 0 \\ 0 & 0 & 0 & 0 & 0 \\ e_2(4) & 0 & e_1(3,1) & 0 & e_2(2) \\ 0 & 0 & 0 & 0 & 0 \\ 0 & 0 & e_2(4) & 0 & e_1(3) \end{bmatrix} \begin{bmatrix} class \\ classroom \\ enrollment \\ teacher \\ student \end{bmatrix}
$$

A graph has no cycle when any entity is a sink or a source or a transit between sink and source. In this situation we have a lattice (partial order) for which we move from source to sink without the possibility to come back (cycle or loop). Given the relation without cycles.

$$
\begin{bmatrix}
 & class & classroom & enrolment & teacher & student \\
class & 0 & 1 & 1 & 1 & 0 \\
classroom & 0 & 0 & 0 & 0 & 0 \\
enrolment & 0 & 0 & 0 & 0 & 1 \\
teacher & 0 & 0 & 0 & 0 & 0 \\
student & 0 & 0 & 0 & 0 & 0
\end{bmatrix}
$$

or

$$
\begin{bmatrix}
 & class & classroom & enrolment & teacher & student \\
class & e_1 & e_2 & e_2 & e_2 & 0 \\
classroom & 0 & e_1 & 0 & 0 & 0 \\
enrolment & 0 & 0 & e_1 & 0 & e_2 \\
teacher & 0 & 0 & 0 & e_1 & 0 \\
student & 0 & 0 & 0 & 0 & e_1
\end{bmatrix}
$$

We have that class is a source, classroom is a sink, enrollment is a transit, teacher is a sink, and student is a sink. The transit moves from class (source) to student (sink) so we have the lattice (Fig. 7.9).

Fig. 7.9 Lattice in the graph of data base

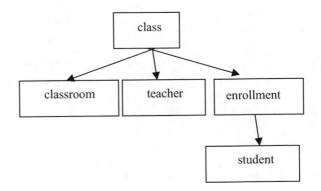

7.5 Graph as a Space with Reference

Given the set of paraboles Fig. 7.10.

We see that in the three cases we respectively have two geometric intersections, one tangent intersection and no intersection. The first situation is given by the equation $x^2 - 1 = 0$. The second situation is given by the equation $x^2 = 0$. And the last situation is given by the equation $x^2 + 1 = 0$. For the last situation, we have no coherent and consistent solutions. In mathematics, given a polynomial equation, when there is no coherent solution, a new unity and coordinate is invented associated with the impossible operation $i = \sqrt{-1}$ that is defined as imaginary coordinate. Now we know that any entity is a unity and has only one name. So we cannot give two different names for one entity. This appears similar to the previous paradox so we solve the paradox in a similar way. We introduce one unity denoted source and the other unity denotes sink. So the same entity is inconsistent because we give two different names at the same entity. The entity can be represented by two dimensional space with two different attributes. The first is source (as e_1) and the other is sink (as e_2). As for the complex number we have $a = x + iy$. For graph theory we have $f = xe_1 + ye_2$.

Where, f is the relationship from the source part of the entity x to the sink part of the entity y. So any entity is split into two parts "sink" as e_1 and "source" as e_2. Now in the complex number we have an algebra of the unities for which we have the multiplication table 7.3.

$$\begin{bmatrix} * & 1 & i \\ 1 & 1 & i \\ i & i & -1 \end{bmatrix} \tag{7.3}$$

So we can define the external and the internal product for the graph coordinate e_1 and e_2.

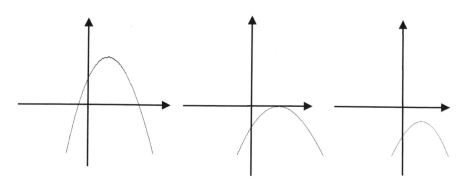

Fig. 7.10 The set of paraboles

$$\text{external product}: \begin{bmatrix} ext & e_i & e_j \\ e_i & 0 & e_i e_j \\ e_j & -e_i e_j & 0 \end{bmatrix}$$

$$\text{internal product}: \begin{bmatrix} int & e_i & e_j \\ e_i & 1 & 0 \\ e_j & 0 & 1 \end{bmatrix}$$

Now we give other possible operations.

Example 7.1
Given the Fig. 7.11

$$\text{general} \begin{bmatrix} * & e_1 & e_2 & \cdots & e_n \\ e_1 & a_{11} & a_{12} & \cdots & a_{1n} \\ e_2 & a_{21} & a_{22} & \cdots & a_{2n} \\ \cdots & \cdots & \cdots & \cdots & \cdots \\ e_n & a_{n1} & a_{n2} & \cdots & a_{nn} \end{bmatrix}$$

Group product

$$\begin{bmatrix} * & e_1 = 1 & e_2 = -1 \\ e_1 = 1 & 1 & -1 \\ e_2 = -1 & -1 & 1 \end{bmatrix}, \begin{bmatrix} * & 1 & -1 & i & -i \\ 1 & 1 & -1 & i & -i \\ -1 & -1 & 1 & -i & i \\ i & i & -i & -1 & 1 \\ -i & -i & i & 1 & -1 \end{bmatrix}$$

where, the relationship is

$$\begin{bmatrix} ae_1 + ae_2 \\ be_1 + be_2 \\ ae_1 + be_2 \\ be_1 + ae_2 \end{bmatrix}$$

In two-dimensional space, there is (Fig. 7.12).

Fig. 7.11 Coherent graph

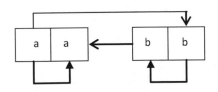

Fig. 7.12 Two-dimensional
representation

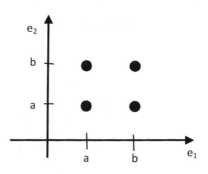

7.6 External Product

Given the relationships

$$\alpha = e_1 v_1 + e_2 v_2$$
$$\beta = e_1 v_2 + e_2 v_3$$

We have

$$\alpha\beta = (e_1 v_1 + e_2 v_2)(e_1 v_2 + e_2 v_3)$$
$$= e_1 e_1 (v_1 v_2) + e_1 e_2 (v_1 v_3) + e_2 e_1 (v_2 v_2) + e_2 e_2 (v_2 v_3)$$

when we use the external product

$$e_i e_j = -e_j e_i$$
$$e_i e_i + e_i e_i = 0$$
$$e_i e_i = 0$$

and

$$\alpha\beta = e_1 e_2 (v_1 v_3) - e_1 e_2 (v_2 v_2) = e_1 e_2 (v_1 v_3 - v_2 v_2)$$

The table of the product is

$$\begin{bmatrix} ext & e_1 v_1 + e_2 v_2 & e_1 v_2 + e_2 v_3 \\ e_1 v_1 + e_2 v_2 & 0 & (v_1 v_3 - v_2 v_2) e_1 e_2 \\ e_1 v_2 + e_2 v_3 & (v_2 v_2 - v_1 v_3) e_1 e_2 & 0 \end{bmatrix}$$

To compute the elements we use the determinant form.

$$\det \begin{bmatrix} v_1 & v_2 \\ v_2 & v_3 \end{bmatrix} = v_1 v_3 - v_2 v_2$$

Determinant provides an easy way to compute the product between relationship. When we permute the column in the determinant we have

$$\det \begin{bmatrix} v_2 & v_1 \\ v_3 & v_2 \end{bmatrix} = v_2 v_2 - v_1 v_3$$

For a chain of three elements we have

$$\alpha\beta\gamma = (e_1 v_1 + e_2 v_2)(e_1 v_2 + e_2 v_3)(e_1 v_3 + e_2 v_4) = 0$$

In fact, for the product, we have almost two equal elements so the product of three is equal to zero. How to explain the result? When e_1 is a query and e_2 is an answer, we say for any answer, there is one and only one question, and for any question, there is one and only one answer. It is impossible to have two queries and only one answer, and also impossible to have two answers with only one query because this has no meaning. This is why the previous product is always false or zero. In the ordinary query and answer system we have no memory because we have only one question at one time and then the answer. We cannot have the memory of two queries (source) and one answer (sink). For the source and sink space we can give a differential form to the superposition of the source sink space and the external product in differential calculus. Given the two relationships for which we have formal entities whose name is a differential operator and the other at the name we have the differential operation for the same function. So

$$\alpha = v_1 e_1 + v_2^1 e_2 = \left(\frac{\partial}{\partial x} e_1 + \frac{\partial}{\partial y} e_2 \right)$$

$$\beta = v_2^2 e_1 + v_3 e_2 = \left(\frac{\partial f}{\partial x} e_1 + \frac{\partial f}{\partial y} e_2 \right)$$

So the external product of two relationship is

$$\alpha\beta = (v_1 e_1 + v_2^1 e_2)(v_2^2 e_1 + v_3 e_2) = e_1 e_2 (v_1 v_3 - v_2^1 v_2^2)$$

$$= e_1 e_2 \left(\frac{\partial}{\partial x} \frac{\partial f}{\partial y} - \frac{\partial}{\partial y} \frac{\partial f}{\partial x} \right)$$

The parallel condition is given by the expression

$$\alpha\beta = (v_1 e_1 + v_2^1 e_2)(v_2^2 e_1 + v_3 e_2) = e_1 e_2 (v_1 v_3 - v_2^1 v_2^2)$$

$$= e_1 e_2 \left(\frac{\partial}{\partial x} \frac{\partial f}{\partial y} - \frac{\partial}{\partial y} \frac{\partial f}{\partial x} \right) = 0$$

We know that the identity $\left(\frac{\partial}{\partial x} \frac{\partial f}{\partial y} - \frac{\partial}{\partial y} \frac{\partial f}{\partial x} \right) = 0$ or parallel condition which can be written also in this way

$$\frac{\partial}{\partial x} \left(\frac{\partial f(x, y)}{\partial y} \right) = \frac{\partial}{\partial y} \left(\frac{\partial f(x, y)}{\partial x} \right)$$

We remark that if the previous coherent condition is true we have the famous integral property (exact differentiable form)

$$df = \frac{\partial f(x, y)}{\partial y} dx + \frac{\partial f(x, y)}{\partial x} dy$$

In fact we have

$$f = \int df = \int \frac{\partial f(x, y)}{\partial x} dx + \int \frac{\partial f(x, y)}{\partial y} dy$$

and

$$\frac{\partial f(x, y)}{\partial y} = \frac{\partial}{\partial y} \left(\int \frac{\partial f(x, y)}{\partial x} dx + \int \frac{\partial f(x, y)}{\partial y} dy \right) = \int \frac{\partial}{\partial y} \frac{\partial f(x, y)}{\partial x} dx + \int \frac{\partial}{\partial y} \frac{\partial f(x, y)}{\partial y} dy$$

$$\frac{\partial f(x, y)}{\partial x} = \frac{\partial}{\partial x} \left(\int \frac{\partial f(x, y)}{\partial x} dx + \int \frac{\partial f(x, y)}{\partial y} dy \right) = \int \frac{\partial}{\partial x} \frac{\partial f(x, y)}{\partial x} dx + \int \frac{\partial}{\partial x} \frac{\partial f(x, y)}{\partial y} dy$$

If

$$\frac{\partial}{\partial x} \frac{\partial f(x, y)}{\partial y} = \frac{\partial}{\partial y} \frac{\partial f(x, y)}{\partial x}$$

then

$$\frac{\partial f(x, y)}{\partial y} = \frac{\partial}{\partial y} \left(\int \frac{\partial f(x, y)}{\partial x} dx + \int \frac{\partial f(x, y)}{\partial y} dy \right) = \int \frac{\partial}{\partial x} \frac{\partial f(x, y)}{\partial y} dx + \int \frac{\partial}{\partial y} \frac{\partial f(x, y)}{\partial y} dy$$

$$\frac{\partial f(x, y)}{\partial x} = \frac{\partial}{\partial x} \left(\int \frac{\partial f(x, y)}{\partial x} dx + \int \frac{\partial f(x, y)}{\partial y} dy \right) = \int \frac{\partial}{\partial x} \frac{\partial f(x, y)}{\partial x} dx + \int \frac{\partial}{\partial y} \frac{\partial f(x, y)}{\partial x} dy$$

That with particular condition is an identity. Now we prove the reverse condition.

If

$$df = \frac{\partial f(x,y)}{\partial x} dx + \frac{\partial f(x,y)}{\partial y} dy$$

then

$$f = \int \frac{\partial f(x,y)}{\partial x} dx + \int \frac{\partial f(x,y)}{\partial y} dy$$

$$\frac{\partial f(x,y)}{\partial y} = \frac{\partial}{\partial y} \left(\int \frac{\partial f(x,y)}{\partial x} dx + \int \frac{\partial f(x,y)}{\partial y} dy \right)$$

$$\frac{\partial f(x,y)}{\partial x} = \frac{\partial}{\partial x} \left(\int \frac{\partial f(x,y)}{\partial x} dx + \int \frac{\partial f(x,y)}{\partial y} dy \right)$$

$$\frac{\partial}{\partial x} \left(\frac{\partial f(x,y)}{\partial y} \right) = \frac{\partial}{\partial x} \left(\frac{\partial}{\partial y} \left(\int \frac{\partial f(x,y)}{\partial x} dx + \int \frac{\partial f(x,y)}{\partial y} dy \right) \right)$$

$$\frac{\partial}{\partial y} \left(\frac{\partial f(x,y)}{\partial x} \right) = \frac{\partial}{\partial y} \left(\frac{\partial}{\partial x} \left(\int \frac{\partial f(x,y)}{\partial x} dx + \int \frac{\partial f(x,y)}{\partial y} dy \right) \right)$$

and

$$\frac{\partial}{\partial x} \left(\frac{\partial f(x,y)}{\partial y} \right) = \frac{\partial}{\partial y} \left(\frac{\partial f(x,y)}{\partial x} \right)$$

The data base form

$$v_2^2 e_1 + v_3 e_2 = \left(\frac{\partial f}{\partial x} e_1 + \frac{\partial f}{\partial y} e_2 \right)$$

is denoted gradient. The divergence of a vector function indicates how much of the field flows outward from a given point. Figure 7.13a shows a function that has divergence. Note that the divergence of a vector field is itself a scalar. If the vector field is a velocity field then a positive divergence implies the mass at the point

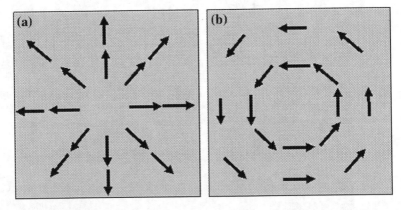

Fig. 7.13 Derivatives of vector functions. **a** An irrotational vector field has only divergence (*no curl*). **b** A solenoidal vector field has only curl (*no divergence*)

decreases. Think about a tank of compressed gas emptying out; the volume of the container remains constant but the amount of gas inside the tank diminishes as gas flows outward.

7.7   Internal Product for Multidimensional Many Sources, and Many Sinks Space

Given the internal product

$$if \ i = j \ then \ e_i e_j = 1$$
$$if \ i \neq j \ then \ e_i e_j = 0$$

we have

$$\alpha\beta = \left(\frac{\partial}{\partial x}e_1 + \frac{\partial}{\partial y}e_2\right)\left(\frac{\partial f}{\partial x}e_1 + \frac{\partial f}{\partial y}e_2\right) = \frac{\partial}{\partial x}\frac{\partial f}{\partial x} + \frac{\partial}{\partial y}\frac{\partial f}{\partial y}$$

The product is the divergence. The curl of a vector field indicates the amount of circulation about each point. Figure 7.13b shows a vector field that has curl. The curl of a velocity field is called the vorticity. Note that the curl is itself a vector. To find its direction, we use the "right-hand rule": Curl the fingers of your right hand along the direction of the vectors and your thumb will point along the direction of the curl. In Fig. 7.13b, the curl points out of the page. The fundamental theorem of vector calculus states that you can represent a vector field as the sum of an irrotational part (which has no curl) and a solenoidal part (which has no divergence).

7.8   Segment Type, Surface Type, Volume Type and Others in Graphs

We know that the area of the parallelogram generated by two vectors is given by the product of the norm of one vector and the distance of the perpendicular line of the other vector. It is shown in Fig. 7.14.

We can consider a surface as a measure of *independence* between vectors. When the vectors are independent, they are orthogonal and the surface assumes the maximum value. Given the vector v, its orthogonal vector is

$$w = (I - v(v^T v)^{-1} v^T) = I - Q(v)$$

and

$$wv = (I - v(v^T v)^{-1} v^T)v = 0$$

Fig. 7.14 The area of the
parallelogram generated by
two vectors

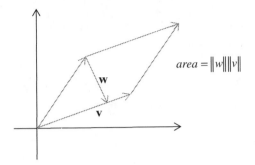

In fact with a little algebra we have

$$wv = (I - v(v^T v)^{-1} v^T)v = v - v(v^T v)^{-1} v^T v = 0$$

To compute the surface generated by the two general vectors (Fig. 7.15) we must have an algorithm that creates the orthogonal projection of one vector into the other. Now we know that the operator

$$Q = v_1 (v_1^T v_1)^{-1} v_1^T$$

Project v_2 into v_1 and we have

$$Qv_2 = v_1 (v_1^T v_1)^{-1} v_1^T v_2$$

With the property

$$Qv_1 = v_1 (v_1^T v_1)^{-1} v_1^T v_1 = v_1$$

The projection of v_1 on itself is equal to v_1. We have also that

$$Q(Qv_2) = v_1 (v_1^T v_1)^{-1} v_1^T v_1 (v_1^T v_1)^{-1} v_1^T v_2 = v_1 (v_1^T v_1)^{-1} v_1^T v_2 = Qv_2$$

The projection of the projection is again the same projection. Now we can see that

Fig. 7.15 Operational form
of the projection operator

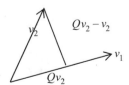

$$(Qv_2 - v_2)^T Qv_2 = ((Qv_2)^T - v_2^T)Qv_2 = (v_2^T Q^T Qv_2 - v_2^T Qv_2)$$

but

$$Q^T = Q$$

and

$$v_2^T Q^T Qv_2 - v_2^T Qv_2 = v_2^T QQv_2 - v_2^T Qv_2 = v_2^T Qv_2 - v_2^T Qv_2 = 0$$

Example 7.2

$$v_1 = \begin{bmatrix} a_1 \\ a_2 \end{bmatrix}, v_2 = \begin{bmatrix} b_1 \\ b_2 \end{bmatrix}, base = \sqrt{\begin{bmatrix} a_1 \\ a_2 \end{bmatrix}^T \begin{bmatrix} a_1 \\ a_2 \end{bmatrix}} = \sqrt{a_1^2 + a_2^2}$$

$$Qv_2 - v_2 = \frac{1}{a_1^2 + a_2^2} \begin{bmatrix} a_2(a_1 b_2 - b_1 a_2) \\ -a_1(a_1 b_2 - b_1 a_2) \end{bmatrix}, (Qv_2 - v_2)^T(Qv_2 - v_2) = \frac{(a_1 b_2 - b_1 a_2)^2}{a_1^2 + a_2^2}$$

$$height = [Qv_2 - v_2)^T(Qv_2 - v_2)]^{1/2} = \frac{(a_1 b_2 - b_1 a_2)}{\sqrt{a_1^2 + a_2^2}} = copula$$

$$surface = height * base = a_1 b_2 - b_1 a_2 = \det \begin{bmatrix} a_1 & b_1 \\ a_2 & b_2 \end{bmatrix}$$

The copula has the geometric image as height but gives us the degree of independence. Because

$$v_2 - Q(v_1)v_2 = (I - Q(v_1))v_2$$

and

$$\begin{aligned} D &= [(v_2 - Q(v_1))v_2]^T[(v_2 - Q(v_1))v_2] \\ &= v_2^T - v_2^T(Q(v_1)v_2) - (Q(v_1)v_2)^T v_2 + (Q(v_1)v_2)^T Q(v_1)v_2 \\ &= v_2^T - v_2^T(Q(v_1)v_2) - (Q(v_1)v_2)^T v_2 + v_2^T(Q(v_1))v_2 \\ &= v_2^T - (Q(v_1)v_2)^T v_2 = v_2^T - v_2^T Q(v_1)v_2 = v_2^T(I - Q(v_1))v_2 \end{aligned}$$

From the previous expression we can have the example

$$Q(v_1)v_2 = \begin{bmatrix} \cos^2(\alpha) & \cos(\alpha)\sin(\alpha) \\ \cos(\alpha)\sin(\alpha) & \sin^2(\alpha) \end{bmatrix} \begin{bmatrix} \cos(\beta) \\ \sin(\beta) \end{bmatrix}$$

$$I - Q(v_1) = \begin{bmatrix} 1 & 0 \\ 0 & 1 \end{bmatrix} - \begin{bmatrix} \cos^2(\alpha) & \cos(\alpha)\sin(\alpha) \\ \cos(\alpha)\sin(\alpha) & \sin^2(\alpha) \end{bmatrix}$$

$$= \begin{bmatrix} \sin^2(\alpha) & \cos(\alpha)\sin(\alpha) \\ \cos(\alpha)\sin(\alpha) & \cos^2(\alpha) \end{bmatrix}$$

and

$$D = v_2^T(I - Q(v_1))v_2$$
$$= \begin{bmatrix} \cos(\beta) \\ \sin(\beta) \end{bmatrix}^T \begin{bmatrix} 1 & 0 \\ 0 & 1 \end{bmatrix} - \begin{bmatrix} \cos^2(\alpha) & \cos(\alpha)\sin(\alpha) \\ \cos(\alpha)\sin(\alpha) & \sin^2(\alpha) \end{bmatrix} \begin{bmatrix} \cos(\beta) \\ \sin(\beta) \end{bmatrix}$$
$$= \sin^2(\alpha - \beta)$$

Given the product operator

$$\alpha\beta = (e_1 v_1 + e_2 v_2)(e_1 v_3 + e_2 v_4) = (v_1 v_4 - v_2 v_3)e_1 e_2$$
$$= \det \begin{bmatrix} v_1 & v_2 \\ v_3 & v_4 \end{bmatrix} e_1 e_2 = (v_1 v_4 - v_2 v_3)e_1 e_2 = D e_1 e_2$$

In fact we have

$$D^2 = \left(\det \begin{bmatrix} \cos(\alpha) & \sin(\alpha) \\ \cos(\beta) & \sin(\beta) \end{bmatrix} \right)^2 = (\cos(\alpha)\sin(\beta) - \sin(\alpha)\cos(\beta))^2$$
$$= \sin^2(\alpha - \beta)$$

For the *DEPENDENCE* we have the expression

$$Dependence = (Qv_2)^T(Qv_2)$$

Example 7.3

Given $v_1 = \begin{bmatrix} \cos(\alpha) \\ \sin(\alpha) \end{bmatrix}, v_2 = \begin{bmatrix} \cos(\beta) \\ \sin(\beta) \end{bmatrix}$

We have the dependence operator E in this way

$$Q(v_1) = \begin{bmatrix} \cos(\alpha) \\ \sin(\alpha) \end{bmatrix} \left(\begin{bmatrix} \cos(\alpha) \\ \sin(\alpha) \end{bmatrix}^T \begin{bmatrix} \cos(\alpha) \\ \sin(\alpha) \end{bmatrix} \right)^{-1} \begin{bmatrix} \cos(\alpha) \\ \sin(\alpha) \end{bmatrix}^T$$
$$= \begin{bmatrix} \cos^2(\alpha) & \cos(\alpha)\sin(\alpha) \\ \cos(\alpha)\sin(\alpha) & \sin^2(\alpha) \end{bmatrix}$$
$$Q(v_1)v_2 = \begin{bmatrix} \cos^2(\alpha) & \cos(\alpha)\sin(\alpha) \\ \cos(\alpha)\sin(\alpha) & \sin^2(\alpha) \end{bmatrix} \begin{bmatrix} \cos(\beta) \\ \sin(\beta) \end{bmatrix}$$
$$= \cos(\alpha - \beta) \begin{bmatrix} \cos(\beta) \\ \sin(\beta) \end{bmatrix}$$
$$E = Dependence = (Q(v_1)v_2)^T Q(v_1)v_2 = v_2^T Q(v_1)^T Q(v_1)v_2$$
$$= v_2^T Q(v_1)Q(v_1)v_2 = v_2^T Q(v_1)v_2$$
$$= \cos^2(\alpha - \beta)$$

For the dependence and independence we can give more information in this way. Given

$$E = Dependence = v_2^T Q(v_1) v_2 = b$$
$$D = Independence = v_2^T (I - Q(v_1)) v_2 = h$$

Vectors in two dimensions

$$Q(v_1) = \begin{bmatrix} a_1 \\ a_2 \end{bmatrix} \left(\begin{bmatrix} a_1 \\ a_2 \end{bmatrix}^T \begin{bmatrix} a_1 \\ a_2 \end{bmatrix} \right)^{-1} \begin{bmatrix} a_1 \\ a_2 \end{bmatrix}^T = \begin{bmatrix} a_1^2 & a_1 a_2 \\ a_1 a_2 & a_2^2 \end{bmatrix}$$

and for $v_2 = \begin{bmatrix} b_1 \\ b_2 \end{bmatrix}$ we have

$$v_2^T Q(v_1) v_2 = \frac{(a_1 b_1 + a_2 b_2)^2}{a_1^2 + a_2^2} = \frac{(a^T b)^2}{a^T a}$$

So

$$\sqrt{v_2^T Q(v_1) v_2} = \frac{(a_1 b_1 + a_2 b_2)^2}{a_1^2 + a_2^2} = \frac{a_1 b_1 + a_2 b_2}{\sqrt{a_1^2 + a_2^2}} = \frac{v_1^T v_2}{\sqrt{a_1^2 + a_2^2}}$$
$$= \frac{\sqrt{a_1^2 + a_2^2} \sqrt{b_1^2 + b_2^2} \cos(\alpha)}{\sqrt{a_1^2 + a_2^2}} = \sqrt{b_1^2 + b_2^2} \cos(\alpha)$$

So we have

$$\frac{\sqrt{v_2^T Q(v_1) v_2}}{\sqrt{b_1^2 + b_2^2}} = \cos(\alpha)$$

and

$$\left(\sqrt{v_2^T Q(v_1) v_2} \right) \sqrt{a_1^2 + a_2^2} = a_1 b_1 + a_2 b_2 = v_1^T v_2$$

For the scalar product

$$v_2^T (I - Q(v_1)) v_2 = \begin{bmatrix} X \\ Y \end{bmatrix}^T \left(\begin{bmatrix} 1 & 0 \\ 0 & 1 \end{bmatrix} - \begin{bmatrix} p_1 \\ p_2 \end{bmatrix} \left(\begin{bmatrix} p_1 \\ p_2 \end{bmatrix}^T \begin{bmatrix} p_1 \\ p_2 \end{bmatrix} \right)^{-1} \begin{bmatrix} p_1 \\ p_2 \end{bmatrix}^T \right) \begin{bmatrix} X \\ Y \end{bmatrix}$$

$$= \begin{bmatrix} X \\ Y \end{bmatrix}^T \begin{bmatrix} p_2^2 & -p_1 p_2 \\ -p_1 p_2 & p_1^2 \end{bmatrix} \begin{bmatrix} X \\ Y \end{bmatrix} \frac{1}{p_1^2 + p_2^2} = independence$$

and

$$S^2 = \begin{bmatrix} X \\ Y \end{bmatrix}^T \begin{bmatrix} p_2^2 & -p_1p_2 \\ -p_1p_2 & p_1^2 \end{bmatrix} \begin{bmatrix} X \\ Y \end{bmatrix} = \sum_{i,j} g_{i,j} X^i X^j$$

$$dS^2 = \begin{bmatrix} dx \\ dy \end{bmatrix}^T \begin{bmatrix} p_2^2 & -p_1p_2 \\ -p_1p_2 & p_1^2 \end{bmatrix} \begin{bmatrix} dx \\ dy \end{bmatrix} = \sum_{i,j} g_{i,j} dx^i dy^j \, surface \; geodesic$$

$$v_2^T Q(v_1) v_2 = \frac{(Xp_1 + Yp_2)^2}{p_1^2 + p_2^2} = dependence.$$

and for the orthogonal projection we have

$$\sqrt{a_1^2 + a_2^2} \sqrt{v_2^T (I - Q(v_1)) v_2} = b_2 a_1 - a_2 b_1 = S = Surface$$

7.9 Orthogonality

Given the vector v, its set of orthogonal vectors are

$$w = (I - v(v^T v)^{-1} v^T) = I - Q(v)$$

and

$$wv = (I - v(v^T v)^{-1} v^T) v = 0$$

When

$$v = \begin{bmatrix} a_1 \\ a_2 \end{bmatrix}$$

$$w = (I - v(v^T v)^{-1} v^T) = \frac{1}{a_1^2 + a_2^2} \begin{bmatrix} a_2^2 & -a_1 a_2 \\ -a_1 a_2 & a_1^2 \end{bmatrix}$$

$$= \frac{1}{a_1^2 + a_2^2} \begin{bmatrix} a_2 \begin{bmatrix} a_2 \\ -a_1 \end{bmatrix} & a_1 \begin{bmatrix} -a_2 \\ a_1 \end{bmatrix} \end{bmatrix}$$

Now at the vector v are associate two orthogonal vectors (columns of w). The surfaces for the vectors

$$\begin{bmatrix} a_2 \\ -a_1 \end{bmatrix} \begin{bmatrix} -a_2 \\ a_1 \end{bmatrix} \; orthogonal \; to \; \begin{bmatrix} a_1 \\ a_2 \end{bmatrix} \; are$$

$$S_1 = \det \begin{bmatrix} a_1 & a_2 \\ a_2 & -a_1 \end{bmatrix} = -(a_1^2 + a_2^2)$$

$$S_2 = \det \begin{bmatrix} a_1 & a_2 \\ -a_2 & a_1 \end{bmatrix} = a_1^2 + a_2^2$$

In fact we have

$$S = (base)(height) = (\sqrt{a_1^2 + a_2^2})\sqrt{(-a_2)^2 + a_1^2} = a_1^2 + a_2^2$$

The orthogonal vectors are independent from the original vector. The surface changes when we change the direction of the orthogonal vector. In a graphic way we have (Fig. 7.16).

Example 7.4

The two orthogonal vectors

$$\begin{bmatrix} a_1 \\ a_2 \end{bmatrix} = \begin{bmatrix} \cos(\alpha) \\ \sin(\alpha) \end{bmatrix} \quad \begin{bmatrix} -a_2 \\ a_1 \end{bmatrix} = \begin{bmatrix} -\sin(\alpha) \\ \cos(\alpha) \end{bmatrix}$$

are the radius vector and tangent vector to a circle (Fig. 7.17).

Given the relationship (segment type in database)

$$\alpha = e_1 v_1 + e_2 v_2$$
$$\beta = e_1 v_3 + e_2 v_4$$

We represent them in Fig. 7.18.
And

$$\alpha\beta = e_1 e_2 (v_1 v_4) - e_1 e_2 (v_3 v_2) = e_1 e_2 (v_1 v_4 - v_3 v_2)$$

By the coordinate form we have Fig. 7.19.

The product $e_1 e_2$ is a new dimension in the database whose entity value $v_1 v_4 - v_2 v_3$ is a surface type value in the database. So we have three dimensions. This is motivation for which we denote external graph product analogous to with the external product in the differential forms that increase the dimension of the space.

Fig. 7.16 Maximum are for orthogonal vectors

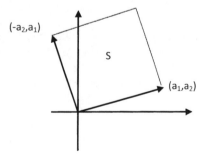

Fig. 7.17 Orthogonal vector
as tangent to the cycle

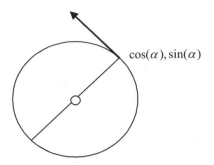

$\cos(\alpha), \sin(\alpha)$

Fig. 7.18 The relationship

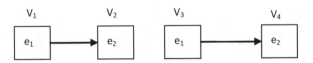

Fig. 7.19 Product of source
sink elements or states

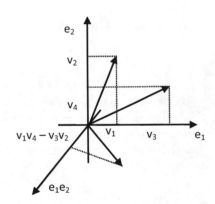

The new dimension makes two relationship comparable. When the two relation-ships are equal, the coordinate e_1e_2 assumes value zero. When the relationships are orthogonal, the new coordinate assumes maximum value. In Fig. 7.20 we show how to compare two relationships in this way.

When the two relationships are adjacent we have

$$\alpha\beta = e_1e_2(v_1v_4) - e_1e_2(v_2v_2) = e_1e_2(v_1v_4 - v_2v_2)$$

Fig. 7.20 The comparison of
the two relationships

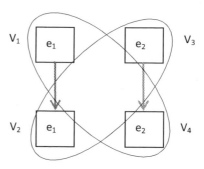

When the two relationship form a loop we have

$$\alpha\beta = e_1e_2(v_1v_1) - e_1e_2(v_2v_2) = e_1e_2(v_1v_1 - v_2v_2)$$

7.10 Multidimensional Graph Space

Given one source point in a graph, from it we move to n sinks. In this situation we
have the space

$$\alpha = e_1v_1 + e_2v_2 + e_3v_3 + \cdots + e_3v_{n+1}$$

where α is a n bifurcation system. The bifurcation gives us the possibility to
increase input and output of the two dimension into output e_1 and inputs e_j of many
dimensions (Fig. 7.21).

Fig. 7.21 Inputs and outputs
of many dimensions

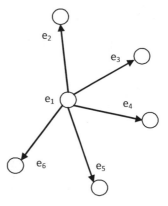

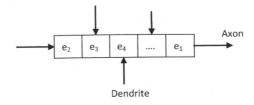

Fig. 7.22 Space of multiple sinks

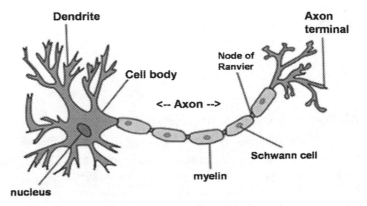

Fig. 7.23 Many sinks (dendrite) many sources (Axon) in neuron model

We can also represent it in Fig. 7.22 with entity.
Bifurcation system is analogous to neuron system (Fig. 7.23).
We show the external product in three dimensional space, Given the elements

$$\alpha = e_1 v_1 + e_2 v_2 + e_3 v_3$$
$$\beta = e_1 v_2 + e_2 v_3 + e_3 v_4$$
$$\gamma = e_1 v_3 + e_2 v_4 + e_3 v_5$$

The product by two elements is

$$\alpha\beta = (e_1 v_1 + e_2 v_2 + e_3 v_3)(e_1 v_4 + e_2 v_5 + e_3 v_6)$$
$$= e_1 e_2 v_1 v_5 + e_1 e_3 v_1 v_6 + e_2 e_1 v_2 v_6 + e_2 e_3 v_2 v_4 + e_3 e_1 v_3 v_4 + e_3 e_2 v_3 v_5$$
$$= e_1 e_2 v_1 v_5 + e_1 e_3 v_1 v_6 - e_1 e_2 v_2 v_6 + e_2 e_3 v_2 v_4 - e_1 e_3 v_3 v_4 - e_2 e_3 v_3 v_5$$
$$= e_1 e_2 (v_1 v_5 - v_2 v_6) + e_1 e_3 (v_1 v_6 - v_3 v_4) + e_2 e_3 (v_2 v_4 - v_3 v_5)$$
$$= e_1 e_2 \det \begin{bmatrix} v_1 & v_2 \\ v_6 & v_5 \end{bmatrix} + e_1 e_3 \det \begin{bmatrix} v_1 & v_3 \\ v_4 & v_6 \end{bmatrix} + e_2 e_3 \det \begin{bmatrix} v_2 & v_5 \\ v_3 & v_4 \end{bmatrix}$$

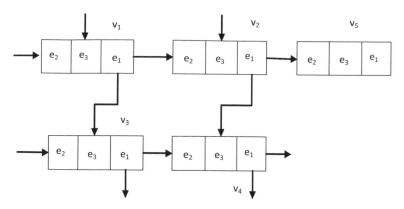

Fig. 7.24 Three dimensional space for sinks and source

The e_2e_3 means the join connection of two answers between two different bifurcations. So e_2v_2 belongs to the first bifurcation and e_3v_4 belongs to the second bifurcation. The product of the two is connection of v_2 as the first answer or e_2 as a query through v_2, e_1 and v_4 as the second answer or e_3. Now the same for v_3 that is the second answer in the first bifurcation and v_5 that is the first answer in the second bifurcation. The two elements are not connected with relationship but are in entanglement situation or synchronic situation (correlate state). We remark that it is impossible to consider the repetition of two questions e_1 or two answers e_2 or two answers e_3. When we split any entity into three parts we have two parts associated with the different inputs (or sinks) and one part for the output (or sources). In Fig. 7.24, we have the two periodic lattice structure.

For three components e_1, e_2, e_3 in the bifurcation, e_1 is the question and e_2 and e_3 are two synchronic or entangled answers. For three dimensions, we have the three product.

$$\alpha\beta\gamma = (e_1v_1 + e_2v_2 + e_3v_3)(e_1v_4 + e_2v_5 + e_3v_6)(e_1v_7 + e_2v_8 + e_3v_9)$$

$$= e_1e_2e_3 \det \begin{bmatrix} v_1 & v_4 & v_7 \\ v_2 & v_5 & v_8 \\ v_3 & v_6 & v_9 \end{bmatrix}$$

Figure 7.25. is the image of the external product in the graph theory.

$$V_1V_2V_3 =$$

$$(V_{11}e_1 + V_{12}e_2 + V_{13}e_3)(V_{21}e_1 + V_{22}e_2 + V_{23}e_3)(V_{31}e_1 + V_{32}e_2 + V_{33}e_3)$$

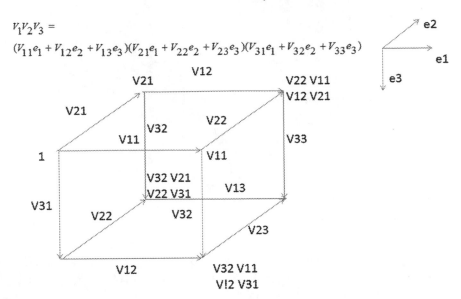

Fig. 7.25   Straight line model by external product

$$a_1 = \begin{bmatrix} x \\ x_1 \\ x_2 \end{bmatrix}, a_2 = \begin{bmatrix} y \\ y_1 \\ y_2 \end{bmatrix}, a_3 = \begin{bmatrix} 1 \\ 1 \\ 1 \end{bmatrix}$$

$$
\begin{aligned}
a_1a_2a_3 &= (xe_1 + x_1e_2 + x_2e_3)(ye_1 + y_1e_2 + y_2e_3)(e_1 + e_2 + e_3) \\
&= (xy)e_1e_1e_1 + (x_1y)e_2e_1e_1 + (x_2y)e_3e_1e_1 + (xy_1)e_1e_2e_1 + (x_1y_1)e_2e_2e_1 \\
&\quad + (x_2y_1)e_3e_2e_1 + (xy_2)e_1e_3e_1 + (x_1y_2)e_2e_3e_1 + (x_2y_2)e_3e_3e_1 + (xy)e_1e_1e_2 \\
&\quad + (x_1y)e_2e_1e_2 + (x_2y)e_3e_1e_2 + (xy_1)e_1e_2e_2 + (x_1y_1)e_2e_2e_2 + (x_2y_1)e_3e_2e_2 \\
&\quad + (xy_2)e_1e_3e_2 + (x_1y_2)e_2e_3e_2 + (x_2y_2)e_3e_3e_2 + (xy)e_1e_1e_3 + (x_1y)e_2e_1e_3 \\
&\quad + (x_2y)e_3e_1e_3 + (xy_1)e_1e_2e_3 + (x_1y_1)e_2e_2e_3 + (x_2y_1)e_3e_2e_3 \\
&\quad + (xy_2)e_1e_3e_3 + (x_1y_2)e_2e_3e_3 + (x_2y_2)e_3e_3e_3
\end{aligned}
$$

and in a more complete way we have

$$
\begin{aligned}
a_1a_2a_3 &= (xe_1 + x_1e_2 + x_2e_3)(ye_1 + y_1e_2 + y_2e_3)(e_1 + e_2 + e_3) \\
&= (x_2y_1)e_3e_2e_1 + (x_1y_2)e_2e_3e_1 + (x_2y)e_3e_1e_2 + (xy_2)e_1e_3e_2 + (x_1y)e_2e_1e_3 \\
&\quad + (xy_1)e_1e_2e_3 = [-(x_2y_1) + (x_1y_2) + (x_2y) - (xy_2) - (x_1y) + (xy_1)]e_1e_2e_3 \\
&= \det \begin{bmatrix} x & y & 1 \\ x_1 & y_1 & 1 \\ x_2 & y_2 & 1 \end{bmatrix} (e_1e_2e_3) = (x(y_1 - y_2) + y(x_2 - x_1) + \det \begin{bmatrix} x_1 & y_1 \\ x_2 & y_2 \end{bmatrix})e_1e_2e_3 = 0
\end{aligned}
$$

or

$$x(y_1 - y_2) + y(x_2 - x_1) + \det\begin{bmatrix} x_1 & y_1 \\ x_2 & y_2 \end{bmatrix} = 0$$

Example for Lie derivative L_X

For special case of the source, sink multi-dimensional space reference we have

$$Xf = X_1 \frac{\partial f}{\partial x_1} + \cdots + X_n \frac{\partial f}{\partial x_n} = (X_1 e_1 + \cdots + X_n e_n)f$$

$$Yf = Y_1 \frac{\partial f}{\partial x_1} + \cdots + Y_n \frac{\partial f}{\partial x_n} = (Y_1 e_1 + \cdots + Y_n e_n)f$$

$$X(Yf) = L_X(Yf) = (X_1 e_1 + \cdots + X_n e_n)[(Y_1 e_1 + \cdots + Y_n e_n)f]$$

$$[(X_1 e_1 + \cdots + X_n e_n)(Y_1 e_1 + \cdots + Y_n e_n)]f +$$

$$(Y_1 e_1 + \cdots + Y_n e_n)[(X_1 e_1 + \cdots + X_n e_n)f] = (L_X Y)f + Y(Xf)$$

$$X(Yf) - Y(Xf) = (L_X Y)f$$

Chapter 8
A New Interpretation of the Determinant as Volume and Entropy

Given the projection operator

$$Qx = ZP(P^T ZP)^{-1}P^T x$$

where

$$x = \begin{bmatrix} x_1 \\ x_2 \\ \cdots \\ x_n \end{bmatrix}$$

$$V_{n+1}^2 = \det \left(\begin{bmatrix} p_{1,1} & \cdots & p_{1,n} & x_1 \\ p_{2,1} & \cdots & p_{2,n} & x_2 \\ p_{3,1} & \cdots & p_{3.n} & x_3 \\ \cdots & \cdots & \cdots & \cdots \\ p_{m,1} & \cdots & p_{m,n} & x_m \end{bmatrix}^T \begin{bmatrix} p_{1,1} & \cdots & p_{1,n} & x_1 \\ p_{2,1} & \cdots & p_{2,n} & x_2 \\ p_{3,1} & \cdots & p_{3.n} & x_3 \\ \cdots & \cdots & \cdots & \cdots \\ p_{m,1} & \cdots & p_{m,n} & x_m \end{bmatrix} \right)$$

$$= \det \left(\begin{bmatrix} \sum_i p_{i,1}^2 & \sum_i p_{i,1}q_{i,2} & \cdots & \sum_i p_{i,1}x_i \\ \sum_i p_{i,1}q_{i,2} & \sum_i p_{i,2}^2 & \cdots & \sum_i p_{i,2}x_i \\ \cdots & \cdots & \cdots & \cdots \\ \sum_i p_{i,1}x_i & \sum_i p_{i,2}x_i & \cdots & \sum_i x_{i,1}^2 \end{bmatrix} \right)$$

When V_n^2 is the square of the n level of volume. When n = 0 we have zero order or point. When n = 1 we have the distance between two points. When n = 2 we have the surface of two dimensions. When n = 3 we have the volume. When n = 4 we have volume in the four dimensional space. At the level n we have volume in the n dimensional space.

© Springer International Publishing AG 2017
G. Resconi et al., *Introduction to Morphogenetic Computing*,
Studies in Computational Intelligence 703, DOI 10.1007/978-3-319-57615-2_8

Example 8.1 Given the set of variables or samples

$$A_i = \begin{bmatrix} p_{i,1} & p_{i,2} & \cdots & x_i \end{bmatrix}$$

The entries in the column vector

$$A = \begin{bmatrix} A_1 \\ A_2 \\ \cdots \\ A_m \end{bmatrix}$$

The volume square matrix is

$$V^2 = A^T A = \sum_{i,j} G(p_{k,h}) x_i x_j$$

For the surface of two dimensions, we have

$$A = \begin{bmatrix} A_1 \\ A_2 \\ \cdots \\ A_m \end{bmatrix} = \begin{bmatrix} p_1 & x_1 \\ p_2 & x_2 \\ \cdots & \cdots \\ p_n & x_n \end{bmatrix}, \quad A^T A = \begin{bmatrix} p_1 & x_1 \\ p_2 & x_2 \\ \cdots & \cdots \\ p_n & x_n \end{bmatrix}^T \begin{bmatrix} p_1 & x_1 \\ p_2 & x_2 \\ \cdots & \cdots \\ p_n & x_n \end{bmatrix} \det \left(\begin{bmatrix} p_1 & x_1 \\ p_2 & x_2 \\ \cdots & \cdots \\ p_n & x_n \end{bmatrix}^T \begin{bmatrix} p_1 & x_1 \\ p_2 & x_2 \\ \cdots & \cdots \\ p_n & x_n \end{bmatrix} \right)$$

$$= \begin{bmatrix} x_1 \\ x_2 \\ \cdots \\ x_n \end{bmatrix}^T \begin{bmatrix} p_2+p_3+\cdots p_n & -p_1p_2 & \cdots & -p_1p_n \\ -p_1p_2 & p_1+p_3+\cdots p_n & \cdots & -p_2p_n \\ \cdots & \cdots & \cdots & \cdots \\ -p_1p_n & -p_2p_n & \cdots & p_1+p_2+\cdots p_{n-1} \end{bmatrix} \begin{bmatrix} x_1 \\ x_2 \\ \cdots \\ x_n \end{bmatrix}$$

So the metric for two dimensional surface square value embedded in n dimension is

$$G(p) = \begin{bmatrix} p_2+p_3+\cdots p_n & -p_1p_2 & \cdots & -p_1p_n \\ -p_1p_2 & p_1+p_3+\cdots p_n & \cdots & -p_2p_n \\ \cdots & \cdots & \cdots & \cdots \\ -p_1p_n & -p_2p_n & \cdots & p_1+p_2+\cdots p_{n-1} \end{bmatrix}$$

And

$S^2 = \sum_{i,j} G(p) x^i x^j$ where S is surface variable embedded in n dimensional space.

Figure 8.1 shows the movement of surface vector.

Figure 8.2 shows the surface vector rotates as an elicoidal system.

We remember that in Euclidean space we have the one dimension metric in n dimensional space.

Fig. 8.1 Volume dynamics

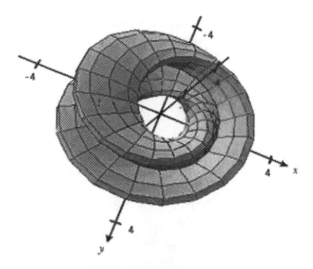

Fig. 8.2 The rotation of the surface in a circle

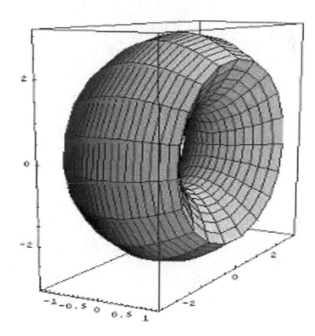

$$A = \begin{bmatrix} A_1 \\ A_2 \\ \cdots \\ A_m \end{bmatrix} = \begin{bmatrix} x_1 \\ x_2 \\ \cdots \\ x_n \end{bmatrix}, \quad A^T A = \det \left(\begin{bmatrix} x_1 \\ x_2 \\ \cdots \\ x_n \end{bmatrix}^T \begin{bmatrix} x_1 \\ x_2 \\ \cdots \\ x_n \end{bmatrix} \right) = x_1^2 + x_2^2 + \cdots + x_n^2 = d^2$$

Now we give a new interpretation to the determinant as volume and also entropy. In fact we have

$$
D = \begin{bmatrix} A_1 - \sum_j A_{1,j} \\ A_2 - \sum_j A_{2,j} \\ \cdots \\ A_m - \sum_j A_{n,j} \end{bmatrix} = \begin{bmatrix} A_1 - \mu_1 \\ A_2 - \mu_2 \\ \cdots \\ A_m - \mu_m \end{bmatrix}
$$

the covariance matrix Σ is the matrix whose (i, j) entry is the covariance

$$
\begin{aligned}
\Sigma_{i,j} &= \operatorname{cov}(A_i, A_j) \\
&= D^T D \\
&= \begin{bmatrix} \sum_j (A_{j1} - \mu_1)(A_{j1} - \mu_1) & \sum_j (A_{j1} - \mu_1)(A_{j2} - \mu_2) & \cdots & \sum_j (A_{jn} - \mu_1)(A_{jn} - \mu_n) \\ \sum_j (A_{j1} - \mu_1)(A_{j2} - \mu_2) & \sum_j (A_{j1} - \mu_1)(A_{j1} - \mu_1) & \cdots & \sum_j (A_{jn} - \mu_1)(A_{jn} - \mu_n) \\ \cdots & \cdots & \cdots & \cdots \\ \sum_j (A_{jn} - \mu_1)(A_{jn} - \mu_n) & \sum_j (A_{jn} - \mu_1)(A_{jn} - \mu_n) & \cdots & \sum_j (A_{j1} - \mu_1)(A_{j1} - \mu_1) \end{bmatrix}
\end{aligned}
$$

The inverse of this matrix, $\sum^{-1}$ is the inverse covariance matrix, also known as the concentration matrix or precision matrix. Differential entropy and log determinant of the covariance matrix of a multivariate Gaussian distribution have many applications in coding, communications, signal processing and statistical inference. We consider in the high dimensional setting optimal estimation of the differential entropy and the log-determinant of the covariance matrix. We first establish a central limit theorem for the log determinant of the sample covariance matrix in the high dimensional setting where the dimension p(n) can grow with the sample size n. An estimator of the differential entropy and the log determinant is then considered. We remember that the determinant of the covariant metric is a volume so we can associate the volume with entropy by log operation.

Proposition *For the orthogonal projection we have*

$$
x^T (I - Q(p_1, \ldots, p_n))x = \frac{V_{n+1}^2}{V_n^2}
$$

Proof For

$$Q(p) = p(p^T p)^{-1} p^T, \quad p = \begin{bmatrix} p_1 \\ p_2 \end{bmatrix}, \quad x = \begin{bmatrix} x_1 \\ x_2 \end{bmatrix}$$

$$V_2^2 = \begin{bmatrix} x_1 \\ x_2 \end{bmatrix} \left(\begin{bmatrix} 1 & 0 \\ 0 & 1 \end{bmatrix} - \begin{bmatrix} p_1 \\ p_2 \end{bmatrix} \left(\begin{bmatrix} p_1 \\ p_2 \end{bmatrix}^T \begin{bmatrix} p_1 \\ p_2 \end{bmatrix} \right)^{-1} \begin{bmatrix} p_1 \\ p_2 \end{bmatrix}^T \right) \begin{bmatrix} x_1 \\ x_2 \end{bmatrix} V_1^2$$

$$= \frac{(p_1 x_2 - p_2 x_1)^2}{p_1^2 + p_2^2} V_1^2$$

but

$$V_1^2 = \begin{bmatrix} p_1 \\ p_2 \end{bmatrix}^T \begin{bmatrix} p_1 \\ p_2 \end{bmatrix} = p_1^2 + p_2^2$$

So

$$V_2^2 = (p_1 x_2 - p_2 x_1)^2$$

Example 8.2

$$\det \left(\begin{bmatrix} p_1 & x_1 \\ p_2 & x_2 \end{bmatrix}^T \begin{bmatrix} p_1 & x_1 \\ p_2 & x_2 \end{bmatrix} \right) = (p_1 x_2 - p_2 x_1)^2 = V_2^2$$

For Fig. 8.3, we can compute the projection operator by volume in the multi-dimensional space.

for

$$V_{n+1}^2 = [x^T (I - Q(p)) x] V_n^2$$

$$x^T x - x^T Q(p) x = \frac{V_{n+1}^2}{V_n^2}$$

and

$$x^T Q(p) x = x^T x - \frac{V_{n+1}^2}{V_n^2} = \frac{(x^T x) V_n^2 - V_{n+1}^2}{V_n^2}$$

where

Fig. 8.3 The projection operator can be computed by volume

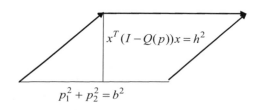

$$x^T (I - Q(p)) x = h^2$$

$$p_1^2 + p_2^2 = b^2$$

$$V_{n+1} = \det(A) = \det\left(\begin{bmatrix} p_{11} & p_{12} & \cdots & x_1 \\ p_{21} & p_{22} & \cdots & x_2 \\ p_{31} & p_{32} & \cdots & x_3 \\ \cdots & \cdots & \cdots & \cdots \\ p_{n1} & p_{n2} & \cdots & x_m \end{bmatrix}\right)$$

where det is the symbol for determinant of a matrix A. When the matrix is a quadratic matrix we have the simple expression.

$$V_{n+1}^2 = \det\left(\begin{bmatrix} p_{11} & p_{12} & \cdots & x_1 \\ p_{21} & p_{22} & \cdots & x_2 \\ p_{31} & p_{32} & \cdots & x_3 \\ \cdots & \cdots & \cdots & \cdots \\ p_{n1} & p_{n2} & \cdots & x_m \end{bmatrix}\right) \det\left(\begin{bmatrix} p_{11} & p_{12} & \cdots & x_1 \\ p_{21} & p_{22} & \cdots & x_2 \\ p_{31} & p_{32} & \cdots & x_3 \\ \cdots & \cdots & \cdots & \cdots \\ p_{n1} & p_{n2} & \cdots & x_m \end{bmatrix}\right)$$

When we have a rectangular matrix we have

$$V_{n+1}^2 = \sum_j (\det A_j)^2$$

where A_j is the quadratic minor of the matrix A. When we compute the quadratic form $x^T Q(p)x$ we can found the matrix of coefficients or projection operator $Q(p)$.

Example 8.3 is for the computation of the projection operator.

$$A = \begin{bmatrix} p_1 \\ p_2 \\ p_3 \end{bmatrix}$$

quadratic minors of A

$A_1 = p_1, A_2 = p_2, A_3 = p_3$

So

$$V_1^2 = [\det(A_1)]^2 + [\det(A_2)]^2 + [\det(A_3)]^2$$
$$= p_1^2 + p_2^2 + p_3^2$$

$$Q(p) = p(p^T p)^{-1} p^T$$

$$p = \begin{bmatrix} p_1 \\ p_2 \\ p_3 \end{bmatrix}, \quad x = \begin{bmatrix} x_1 \\ x_2 \\ x_3 \end{bmatrix}, \quad A = \begin{bmatrix} p_1 & x_1 \\ p_2 & x_2 \\ p_3 & x_3 \end{bmatrix}$$

quadratic minors of A

$$A_1 = \begin{bmatrix} p_1 & x_1 \\ p_2 & x_2 \end{bmatrix}, \quad A_2 = \begin{bmatrix} p_1 & x_1 \\ p_3 & x_3 \end{bmatrix}, \quad A_3 = \begin{bmatrix} p_2 & x_2 \\ p_3 & x_3 \end{bmatrix}$$

So

$$V_2^2 = [\det(A_1)]^2 + [\det(A_2)]^2 + [\det(A_3)]^2$$
$$= (p_1 x_2 - p_2 x_1)^2 + (p_1 x_3 - p_3 x_1)^2 + (p_2 x_3 - p_3 x_2)^2$$

$$\det \left(\begin{bmatrix} p_1 & x_1 \\ p_2 & x_2 \\ p_3 & x_3 \end{bmatrix}^T \begin{bmatrix} p_1 & x_1 \\ p_2 & x_2 \\ p_3 & x_3 \end{bmatrix} \right) = \det \begin{bmatrix} p_1^2 + p_2^2 + p_3^2 & p_1 x_1 + p_2 x_2 + p_3 x_3 \\ p_1 x_1 + p_2 x_2 + p_3 x_3 & x_1^2 + x_2^2 + x_3^2 \end{bmatrix}$$
$$= (p_1 x_2 - p_2 x_1)^2 + (p_1 x_3 - p_3 x_1)^2 + (p_2 x_3 - p_3 x_2)^2$$

So we have

$$x^T Q(p) x = x^T p (p^T p)^{-1} p^T x = \frac{V_1^2 x^T x - V_2^2}{V_1^2}$$

$$= \frac{(p_1^2 + p_2^2 + p_3^2)(x_1^2 + x_2^2 + x_3^2) - [(p_1 x_2 - p_2 x_1)^2 + (p_1 x_3 - p_3 x_1)^2 + (p_2 x_3 - p_3 x_2)^2]}{p_1^2 + p_2^2 + p_3^2}$$

$$= \frac{(p_1^2 + p_2^2 + p_3^2)(x_1^2 + x_2^2 + x_3^2) - [(p_1 x_2 - p_2 x_1)^2 + (p_1 x_2 - p_3 x_1)^2 + (p_2 x_3 - p_3 x_2)^2]}{p_1^2 + p_2^2 + p_3^2}$$

$$= \frac{(p_1 x_1 + p_2 x_2 + p_3 x_3)^2}{p_1^2 + p_2^2 + p_3^2}$$

For a direct calculus we have

$$x^T Q(p) x = x^T p (p^T p)^{-1} p^T x$$

$$= \begin{bmatrix} x_1 \\ x_2 \\ x_3 \end{bmatrix}^T \begin{bmatrix} p_1 \\ p_2 \\ p_3 \end{bmatrix} \left(\begin{bmatrix} p_1 \\ p_2 \\ p_3 \end{bmatrix}^T \begin{bmatrix} p_1 \\ p_2 \\ p_3 \end{bmatrix} \right)^{-1} \begin{bmatrix} p_1 \\ p_2 \\ p_3 \end{bmatrix}^T \begin{bmatrix} x_1 \\ x_2 \\ x_3 \end{bmatrix}$$

$$= \frac{(p_1 x_1 + p_2 x_2 + p_3 x_3)^2}{p_1^2 + p_2^2 + p_3^2}$$

Fig. 8.4 Determinant minor
image by a cube

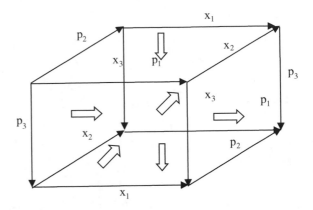

Remark

$$x^T Q(p)x = x^T x - \frac{V_{n+1}^2}{V_n^2} = \frac{(x^T x)V_n^2 - V_{n+1}^2}{V_n^2}$$

The examples above shows that the project of x can be obtained through volumes (V_n *and* V_{n+1}) which is easier to be computed through determinants. We have Fig. 8.4.

We compute the volume V_{n+1} by adding the values of all quadratic minors of the matrix A together. If vectors are three-dimensional, we have three quadratic minors.

$$A_1 = \begin{bmatrix} p_1 & x_1 \\ p_2 & x_2 \end{bmatrix}, \quad A_2 = \begin{bmatrix} p_1 & x_1 \\ p_3 & x_3 \end{bmatrix}, \quad A_3 = \begin{bmatrix} p_2 & x_2 \\ p_3 & x_3 \end{bmatrix},$$

And a cube is formed with six surfaces, but actually only three components because each pair of surface is parallel. We get the value of each surface by calculating its determinant. Since each pair of surface which is parallel has the same value as opposite direction, the sum of the values of all surfaces should be zero. If vectors are two-dimensional, we can get two surfaces with opposite directions and the same value, the sum of the values of all surfaces is also zero.

For the three dimensions we have

$$Q(p, q) = p(p^T p)^{-1} p^T, \quad p = \begin{bmatrix} p_1 & q_1 \\ p_2 & q_2 \\ p_3 & q_3 \end{bmatrix}, \quad x = \begin{bmatrix} x_1 \\ x_2 \\ x_3 \end{bmatrix}$$

$$\begin{bmatrix} x_1 \\ x_2 \\ x_3 \end{bmatrix}^T \begin{pmatrix} \begin{bmatrix} 1 & 0 \\ 0 & 1 \end{bmatrix} - \begin{bmatrix} p_1 & q_1 \\ p_2 & q_2 \\ p_3 & q_3 \end{bmatrix} \begin{pmatrix} \begin{bmatrix} p_1 & q_1 \\ p_2 & q_2 \\ p_3 & q_3 \end{bmatrix}^T \begin{bmatrix} p_1 & q_1 \\ p_2 & q_2 \\ p_3 & q_3 \end{bmatrix} \end{pmatrix}^{-1} \begin{bmatrix} p_1 & q_1 \\ p_2 & q_2 \\ p_3 & q_3 \end{bmatrix}^T \end{pmatrix} \begin{bmatrix} x_1 \\ x_2 \\ x_3 \end{bmatrix} V_2^2.$$

$$= \frac{\det\left(\begin{bmatrix} p_1 & q_1 & x_1 \\ p_2 & q_2 & x_2 \\ p_3 & q_3 & x_3 \end{bmatrix}^T \begin{bmatrix} p_1 & q_1 & x_1 \\ p_2 & q_2 & x_2 \\ p_3 & q_3 & x_3 \end{bmatrix} \right)}{(p_1 x_2 - p_2 x_1)^2 + (p_1 x_2 - p_3 x_1)^2 + (p_2 x_3 - p_3 x_2)^2} V_2^2$$

$$= \det\left(\begin{bmatrix} p_1 & q_1 & x_1 \\ p_2 & q_2 & x_2 \\ p_3 & q_3 & x_3 \end{bmatrix}^T \begin{bmatrix} p_1 & q_1 & x_1 \\ p_2 & q_2 & x_2 \\ p_3 & q_3 & x_3 \end{bmatrix} \right)$$

$$= \det\left(\begin{bmatrix} p_1^2 + p_2^2 + p_3^2 & p_1 q_1 + p_2 q_2 + p_3 q_3 & p_1 x_1 + p_2 x_2 + p_3 x_3 \\ p_1 q_1 + p_2 q_2 + p_3 q_3 & q_1^2 + q_2^2 + q_3^2 & q_1 x_1 + q_2 x_2 + q_3 x_3 \\ p_1 x_1 + p_2 x_2 + p_3 x_3 & q_1 x_1 + q_2 x_2 + q_3 x_3 & x_1^2 + x_2^2 + x_3^2 \end{bmatrix} \right) = V_3^2$$

For non orthogonal projection we have the projection.

$$Q(Z, a) = \begin{bmatrix} Z_{11} & Z_{12} \\ Z_{21} & Z_{22} \end{bmatrix} \begin{bmatrix} a_1 \\ a_2 \end{bmatrix} \left(\begin{bmatrix} a_1 \\ a_2 \end{bmatrix}^T \begin{bmatrix} Z_{11} & Z_{12} \\ Z_{21} & Z_{22} \end{bmatrix} \begin{bmatrix} a_1 \\ a_2 \end{bmatrix} \right)^{-1} \begin{bmatrix} a_1 \\ a_2 \end{bmatrix}^T$$

$$for \quad \begin{bmatrix} b_1 \\ b_2 \end{bmatrix} = \begin{bmatrix} Z_{11} & Z_{12} \\ Z_{21} & Z_{22} \end{bmatrix} \begin{bmatrix} a_1 \\ a_2 \end{bmatrix}$$

$$Q(Z, a) = \begin{bmatrix} Z_{11} & Z_{12} \\ Z_{21} & Z_{22} \end{bmatrix} \begin{bmatrix} a_1 \\ a_2 \end{bmatrix} \left(\begin{bmatrix} a_1 \\ a_2 \end{bmatrix}^T \begin{bmatrix} Z_{11} & Z_{12} \\ Z_{21} & Z_{22} \end{bmatrix} \begin{bmatrix} a_1 \\ a_2 \end{bmatrix} \right)^{-1} \begin{bmatrix} a_1 \\ a_2 \end{bmatrix}^T$$

$$Q(Z, a) = \begin{bmatrix} b_1 \\ b_2 \end{bmatrix} \left(\begin{bmatrix} a_1 \\ a_2 \end{bmatrix}^T \begin{bmatrix} b_1 \\ b_2 \end{bmatrix} \right)^{-1} \begin{bmatrix} a_1 \\ a_2 \end{bmatrix}^T = Q(b, a)$$

Figure 8.5 is the geometric image for two dimensional space.

$$[(I - Q(a, b))x]^T [(I - Q(a, b))x] = [AB]^2$$
$$b^T b = [CB]^2$$
$$[Q(a)b]^T [Q(a)b] = [CD]^2$$
$$[(I - Q(b))x]^T [(I - Q(b))x] = [AE]^2$$

And

$$S^2 = [CD]^2 [AB]^2 = [CB]^2 [AE]^2$$

Fig. 8.5 Area of the triangle by oblique projection

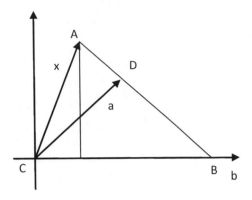

And

$$[AB]^2 = \frac{[CB]^2[AE]^2}{[CD]^2} = [(I - Q(a,b))x]^T[(I - Q(a,b))x]$$

Example

$$[(I - Q(a,b))x]^T[(I - Q(a,b))x]$$

$$\left(\begin{bmatrix} 1 & 0 \\ 0 & 1 \end{bmatrix} - \begin{bmatrix} b_1 \\ b_2 \end{bmatrix} \left(\begin{bmatrix} a_1 \\ a_2 \end{bmatrix}^T \begin{bmatrix} b_1 \\ b_2 \end{bmatrix} \right)^{-1} \begin{bmatrix} a_1 \\ a_2 \end{bmatrix}^T \right)^T$$

$$\left(\begin{bmatrix} 1 & 0 \\ 0 & 1 \end{bmatrix} - \begin{bmatrix} b_1 \\ b_2 \end{bmatrix} \left(\begin{bmatrix} a_1 \\ a_2 \end{bmatrix}^T \begin{bmatrix} b_1 \\ b_2 \end{bmatrix} \right)^{-1} \begin{bmatrix} a_1 \\ a_2 \end{bmatrix}^T \right)$$

$$= \det\left(\begin{bmatrix} b_1 & x_1 \\ b_2 & x_2 \end{bmatrix} \right)^2 \frac{a_1^2 + a_2^2}{(a_1b_1 + a_2b_2)^2} = S^2 \frac{a_1^2 + a_2^2}{(a_1b_1 + a_2b_2)^2}$$

where,

$$\frac{a_1^2 + a_2^2}{(a_1b_1 + a_2b_2)^2}$$

is internal element to the oblique projection operator. Now for the projection operator property (dependence) we have

$$\frac{(a_1b_1 + a_2b_2)^2}{a_1^2 + a_2^2} = \begin{bmatrix} a_1 \\ a_2 \end{bmatrix} \left(\begin{bmatrix} a_1 \\ a_2 \end{bmatrix}^T \begin{bmatrix} a_1 \\ a_2 \end{bmatrix} \right)^{-1} \begin{bmatrix} a_1 \\ a_2 \end{bmatrix}^T \begin{bmatrix} b_1 \\ b_2 \end{bmatrix}$$

$$= [Q(a)b]^T[Q(a)b] = [CD]^2$$

So for OBLIQUE projection we have

$$|(I - Q(a,b))x|^2|(Q(a))b|^2 = \det\left(\begin{bmatrix} b_1 & x_1 \\ b_2 & x_2 \end{bmatrix}\right)^2$$

Now the bifurcation is a three dimensional element. Two different bifurcations are given by the two vectors

$$p = \begin{bmatrix} p_1 \\ p_2 \\ p_3 \end{bmatrix}, \quad x = \begin{bmatrix} x_1 \\ x_2 \\ x_3 \end{bmatrix}$$

And

$$x^T(Q(p))x = \begin{bmatrix} x_1 \\ x_2 \\ x_3 \end{bmatrix}^T \begin{bmatrix} p_1 \\ p_2 \\ p_3 \end{bmatrix} \left(\begin{bmatrix} p_1 \\ p_2 \\ p_3 \end{bmatrix}^T \begin{bmatrix} p_1 \\ p_2 \\ p_3 \end{bmatrix}\right)^{-1} \begin{bmatrix} p_1 \\ p_2 \\ p_3 \end{bmatrix}^T \begin{bmatrix} x_1 \\ x_2 \\ x_3 \end{bmatrix}$$

$$= \begin{bmatrix} x_1 \\ x_2 \\ x_3 \end{bmatrix}^T \begin{bmatrix} p_1^2 & p_1p_2 & p_1p_3 \\ p_1p_2 & p_1^2 & p_2p_3 \\ p_1p_3 & p_2p_3 & p_1^2 \end{bmatrix} \begin{bmatrix} x_1 \\ x_2 \\ x_3 \end{bmatrix} \frac{1}{p_1^2 + p_2^2 + p_3^2}$$

$$= \frac{(x_1p_1 + x_2p_2 + x_3p_3)^2}{p_1^2 + p_2^2 + p_3^2} = \frac{(x^Tp)^2}{p^Tp}$$

$$v_2^T(I - Q(v_1))v_2 = \begin{bmatrix} X \\ Y \\ Z \end{bmatrix}^T \left(\begin{bmatrix} 1 & 0 & 0 \\ 0 & 1 & 0 \\ 0 & 0 & 1 \end{bmatrix} - \begin{bmatrix} p_1 \\ p_2 \\ p_3 \end{bmatrix}\begin{bmatrix} p_1 \\ p_2 \\ p_3 \end{bmatrix}^T\right)\begin{bmatrix} X \\ Y \\ Z \end{bmatrix} \cdot \frac{1}{\det\left[\begin{bmatrix} p_1 \\ p_2 \\ p_3 \end{bmatrix}^T \begin{bmatrix} p_1 \\ p_2 \\ p_3 \end{bmatrix}\right]}$$

$$= \frac{\begin{bmatrix} X \\ Y \\ Z \end{bmatrix}^T \begin{bmatrix} p_1^2 + p_2^2 + p_3^2 - p_1^2 & -p_1p_2 & -p_1p_3 \\ -p_1p_2 & p_1^2 + p_2^2 + p_3^2 - p_2^2 & -p_2p_3 \\ -p_1p_3 & -p_2p_3 & p_1^2 + p_2^2 + p_3^2 - p_3^2 \end{bmatrix}\begin{bmatrix} X \\ Y \\ Z \end{bmatrix}}{p_1^2 + p_2^2 + p_3^2}$$

$$= \frac{(Xp_2 - p_1Y)^2 + (Zp_1 - p_3X)^2 + (Zp_2 - p_3Y)^2}{p_1^2 + p_2^2 + p_3^2}$$

We remark that we can find the projection operator from the determinant.

$$
S^2 = \det \left(\begin{bmatrix} X & p_1 \\ Y & p_2 \\ Z & p_3 \end{bmatrix}^T \begin{bmatrix} X & p_1 \\ Y & p_2 \\ Z & p_3 \end{bmatrix} \right) = (Xp_2 - p_1 Y)^2 + (Zp_1 - p_3 X)^2
$$

$$
+ (Zp_2 - p_3 Y)^2 = \sum_{i,j} g_{i,j} X^i X^j
$$

where the metric g is a parametric metric

$$
g(p_1, p_2, p_3) = \begin{bmatrix} p_2 + p_3 & -p_1 p_2 & -p_1 p_3 \\ -p_1 p_2 & p_1 + p_3 & -p_2 p_3 \\ -p_1 p_3 & -p_2 p_3 & p_2 + p_1 \end{bmatrix}
$$

$$
Volume = \det \left(\begin{bmatrix} x_1 & p_1 & q_1 \\ x_2 & p_2 & q_2 \\ x_3 & p_3 & q_3 \end{bmatrix} \right) = |(I - Q(p,q)x)|^2 (Surface)^2
$$

where

$$
Q(p,q)x = \begin{bmatrix} p_1 & q_1 \\ p_2 & q_2 \\ p_3 & q_3 \end{bmatrix} \left(\begin{bmatrix} p_1 & q_1 \\ p_2 & q_2 \\ p_3 & q_3 \end{bmatrix}^T \begin{bmatrix} p_1 & q_1 \\ p_2 & q_2 \\ p_3 & q_3 \end{bmatrix} \right)^{-1} \begin{bmatrix} p_1 & q_1 \\ p_2 & q_2 \\ p_3 & q_3 \end{bmatrix}^T \begin{bmatrix} x_1 \\ x_2 \\ x_3 \end{bmatrix}
$$

8.1   De Bruijn Graph Evolution by Skew Product

Given the structure (Fig. 8.6)
   Network at low level

$$
M = \begin{bmatrix} & a & b \\ a & 1 & 1 \\ b & 1 & 1 \end{bmatrix},
$$

$$
V = \left[\begin{bmatrix} 1 & 0 \\ 0 & 1 \end{bmatrix} e_1 + \begin{bmatrix} 1 & 1 \\ 1 & 1 \end{bmatrix} e_2 \right] \begin{bmatrix} a \\ b \end{bmatrix} = \begin{bmatrix} a(e_1 + e_2) & be_2 \\ be_2 & b(e_1 + e_2) \end{bmatrix}
$$

   Second level

$$
\begin{bmatrix}
skew & ae_1 + ae_2 & ae_1 + be_2 & be_1 + ae_2 & be_1 + be_2 \\
ae_1 + ae_2 & (ae_1 + ae_2)(ae_1 + ae_2) & (ae_1 + ae_2)(ae_1 + be_2) & (ae_1 + ae_2)(be_1 + ae_2) & (ae_1 + ae_2)(be_1 + be_2) \\
ae_1 + be_2 & (ae_1 + be_2)(ae_1 + ae_2) & (ae_1 + be_2)(ae_1 + be_2) & (ae_1 + be_2)(be_1 + ae_2) & (ae_1 + be_2)(be_1 + be_2) \\
be_1 + ae_2 & (be_1 + ae_2)(ae_1 + ae_2) & (be_1 + ae_2)(ae_1 + be_2) & (be_1 + ae_2)(be_1 + ae_2) & (be_1 + ae_2)(be_1 + be_2) \\
be_1 + be_2 & (be_1 + be_2)(ae_1 + ae_2) & (be_1 + be_2)(ae_1 + be_2) & (be_1 + be_2)(be_1 + ae_2) & (be_1 + be_2)(be_1 + be_2)
\end{bmatrix}
$$

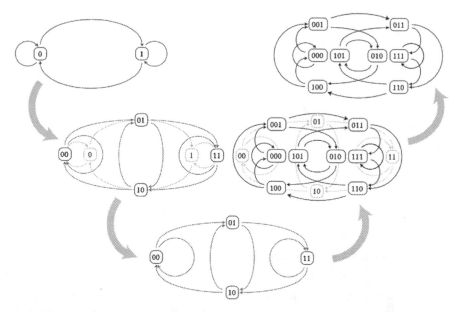

Fig. 8.6 Graph evolution

The SKEW PRODUCT gives us the FILTER that separates parts of the graph in a way to know if two relationships are coupled by a common element (relationships are adjacent) we can also see if two relationships form a cycle and many other properties for composition of chains.

$$
\begin{bmatrix}
skew & ae_1 + ae_2 & ae_1 + be_2 & be_1 + ae_2 & be_1 + be_2 \\
ae_1 + ae_2 & (a^2 - a^2)e_1e_2 & (ab - a^2)e_1e_2 & (a^2 - ab)e_1e_2 & (ab - ab)e_1e_2 \\
ae_1 + be_2 & (a^2 - ab)e_1e_2 & (a^2 - ab)e_1e_2 & (a^2 - b^2)e_1e_2 & (ab - b^2)e_1e_2 \\
be_1 + ae_2 & (ab - a^2)e_1e_2 & (b^2 - a^2)e_1e_2 & (ab - ab)e_1e_2 & (b^2 - ab)e_1e_2 \\
be_1 + be_2 & (ab - ab)e_1e_2 & (b^2 - ab)e_1e_2 & (ab - b^2)e_1e_2 & (b^2 - b^2)e_1e_2
\end{bmatrix}
$$

In the line graph, there is adjacence to 1 and no adjacence to zero. The line graph that connects two different relationships has 8 elements generating from the original 4 elements. See

$$
\begin{bmatrix}
skew & ae_1 + ae_2 & ae_1 + be_2 & be_1 + ae_2 & be_1 + be_2 \\
ae_1 + ae_2 & 1 & 1 & 0 & 0 \\
ae_1 + be_2 & 0 & 0 & 1 & 1 \\
be_1 + ae_2 & 1 & 1 & 0 & 0 \\
be_1 + be_2 & 0 & 0 & 1 & 1
\end{bmatrix}
$$

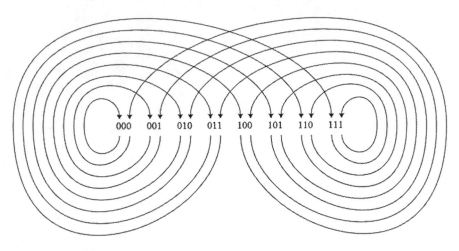

Fig. 8.7 Other graph evolution system

That is the second generation of the initial graph. Now we can repeat again the same process and we can have any type of graph evolution. Another example of evolution is given by Fig. 8.7.

The evolution of the previous graph can be obtained again by the source, sink skew product space.

Chapter 9
Morphogenetic Computing in Genetic Algorithms

9.1 Projection Instrument as Formal Sink Source Change with Invariance

9.1.1 Mophogenetic Transformation

For the multidimensional space S the transformation of vector X in this space is obtained by a quadratic matrix A. So we have (9.1).

$$Y = A\,X \tag{9.1}$$

where Y is the output, X is the input and A is the system (Fig. 9.1).

The output Y is a dependent variable and X is the independent variable or free variable. When A is a quadratic non-singular matrix we can change the direction and Y becomes the independent variable and X the dependent variable. So we have Fig. 9.2.

Or (9.2).

$$X = A^{-1}Y \tag{9.2}$$

The variable Y is an expected output and X is the expected input. For the relation (9.1) and (9.2) we have the loop (9.3).

$$Y = A\,X = A\left[A^{-1}Y\,\right] = Y \tag{9.3}$$

Given the expected input Y we compute the expected input and after we compute the real Y by AX. In the previous case the initial value of Y is equal to the final Y by inverse matrix of the system A. Because the system is a quadratic non-singular matrix the expected output becomes real.

© Springer International Publishing AG 2017
G. Resconi et al., *Introduction to Morphogenetic Computing*,
Studies in Computational Intelligence 703, DOI 10.1007/978-3-319-57615-2_9

Fig. 9.1 The input and
output on system A

Fig. 9.2 The reverse of input
and output system

9.1.2   Inverse Problem in Systems with Different Numbers of Inputs and Outputs

If A is a rectangular matrix the transformation (9.2) has no meaning. For example, given

$$A = \begin{bmatrix} 1 & 0 \\ 1 & 1 \\ 0 & 1 \end{bmatrix}, \quad y = \begin{bmatrix} y_1 \\ y_2 \\ y_3 \end{bmatrix}$$

We have

$$Ax = \begin{bmatrix} 1 & 0 \\ 1 & 1 \\ 0 & 1 \end{bmatrix} \begin{bmatrix} x_1 \\ x_2 \end{bmatrix} = \begin{bmatrix} x_1 \\ x_1 + x_2 \\ x_2 \end{bmatrix} = \begin{bmatrix} y_1 \\ y_2 \\ y_3 \end{bmatrix} \tag{9.4}$$

Form two inputs $x = \begin{bmatrix} x_1 \\ x_2 \end{bmatrix}$ we compute the values of three outputs $y = \begin{bmatrix} y_1 \\ y_2 \\ y_3 \end{bmatrix}$.

In an explicit way we have the system (9.5)

$$\begin{cases} x_1 = y_1 \\ x_1 + x_2 = y_2 \\ x_2 = y_3 \end{cases} \tag{9.5}$$

We try to solve the system by substitution (9.6).

$$y_1 + y_3 = y_2 \tag{9.6}$$

In a graphic way we have Fig. 9.3.

We see that the variables in input disappear and between output variables we have a condition or constraint. So among the three values of the output, only two values are free and the third value is not free.

With Eq. (9.1), we try to build the inverse system of (9.1) when matrix A is rectangular. We move from two dimensions to three dimensions by the up operator A.

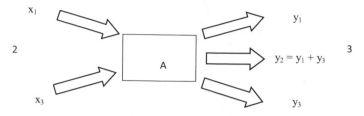

Fig. 9.3 Inputs and outputs with different numbers

Because the inverse of A does not exist, we introduce an operator B and have (9.7).

$$BA = Identity \tag{9.7}$$

One of the possible solutions of Eq. (9.7) where B is the unknown operator or matrix is

$$B = (A^T A)^{-1} A^T \tag{9.8}$$

In fact we have

$$(A^T A)^{-1} A^T A = Identity$$

And for the vector y as source or input we have

$$Qy = ABy = A(A^T A)^{-1} A^T y$$
$$when$$
$$B = A^{-1} \tag{9.9}$$
$$Qy = y$$

For (9.4) and (9.5) system, we begin with $y = \begin{bmatrix} y_1 \\ y_2 \\ y_3 \end{bmatrix}$ as free variable input in three dimensional space. With the operator AT, we move from three-dimensional space to two dimensional space.

$$A^T y = \begin{bmatrix} 1 & 0 \\ 1 & 1 \\ 0 & 1 \end{bmatrix}^T \begin{bmatrix} y_1 \\ y_2 \\ y_3 \end{bmatrix} = \begin{bmatrix} 1 & 1 & 0 \\ 0 & 1 & 1 \end{bmatrix} \begin{bmatrix} y_1 \\ y_2 \\ y_3 \end{bmatrix} = \begin{bmatrix} y_1 + y_2 \\ y_2 + y_3 \end{bmatrix} = A^T Y \tag{9.10}$$

The operator AT A is a 2×2 dimension. In fact for (9.4) and (9.5) we have

$$A^T A = \begin{bmatrix} 1 & 0 \\ 1 & 1 \\ 0 & 1 \end{bmatrix}^T \begin{bmatrix} 1 & 0 \\ 1 & 1 \\ 0 & 1 \end{bmatrix} = \begin{bmatrix} 1 & 1 & 0 \\ 0 & 1 & 1 \end{bmatrix} \begin{bmatrix} 1 & 0 \\ 1 & 1 \\ 0 & 1 \end{bmatrix} = \begin{bmatrix} 2 & 1 \\ 1 & 2 \end{bmatrix} \tag{9.11}$$

And

$$(A^T A)^{-1} A^T y = \begin{bmatrix} 2 & 1 \\ 1 & 2 \end{bmatrix}^{-1} \begin{bmatrix} y_1 + y_2 \\ y_2 + y_3 \end{bmatrix} = \frac{1}{3} \begin{bmatrix} 2 & -1 \\ -1 & 2 \end{bmatrix} \begin{bmatrix} y_1 + y_2 \\ y_2 + y_3 \end{bmatrix}$$

$$= \frac{1}{3} \begin{bmatrix} 2y_1 + y_2 + y_3 \\ y_1 + y_2 + 2y_3 \end{bmatrix} = \begin{bmatrix} x_1 \\ x_2 \end{bmatrix} = x \tag{9.12}$$

In the end, we have the last passage

$$Qy = Ax = \frac{1}{3} \begin{bmatrix} 1 & 0 \\ 1 & 1 \\ 0 & 1 \end{bmatrix} \begin{bmatrix} 2y_1 + y_2 + y_3 \\ y_1 + y_2 + 2y_3 \end{bmatrix} = \frac{1}{3} \begin{bmatrix} 2y_1 + y_2 + y_3 \\ 2y_1 + y_2 + y_3 + y_1 + y_2 + 2y_3 \\ y_1 + y_2 + 2y_3 \end{bmatrix} \tag{9.13}$$

where we combine the components of the vector y in a way to have the invariant form (9.5). In fact, the second component of Qy is the sum of the first plus the third. As a system process we have Fig. 9.4 where we generate a vector Qy from many types of the vectors y for which the invariant is always true. Figure 9.4 shows the complete process and the inverse process that we generate a new y or Qy from x and we come back to x from y, which satisfies the invariant form.

Given the module of the difference between internal and external values in (9.14).

$$D = (y - Ax)^T (y - Ax) \tag{9.14}$$

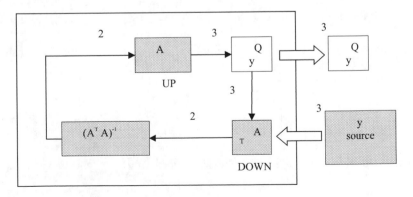

Fig. 9.4 Conceptual filter or morpho filter

Now we choose among all vectors which satisfy invariant form (9.6), the distance between y and Ax has the minimum. So we have

$$D = (y - Ax)^T (y - Ax) = y^T y - y^T Ax - (Ax)^T y + (Ax)^T (Ax) \qquad (9.15)$$

To compute the minimum, we make the derivatives of the previous form

$$\frac{\partial D}{\partial x_j} = -y^T A \frac{\partial x}{\partial x_j} - \frac{\partial x^T}{\partial x_j} A^T y + \frac{\partial x^T}{\partial x_j} A^T Ax + x^T A^T A \frac{\partial x}{\partial x_j} \qquad (9.16)$$

where

$$x = \begin{bmatrix} x_1 \\ \cdots \\ x_j \\ x_{j+1} \\ \cdots \\ x_p \end{bmatrix} \quad \text{and} \quad \frac{\partial x}{\partial x_j} = \begin{bmatrix} 0 \\ \cdots \\ 1 \\ 0 \\ \cdots \\ 0 \end{bmatrix} = v_j, \quad \frac{\partial x^T}{\partial x_j} = [0 \ \cdots \ 1 \ 0 \ \cdots \ 0] = v_j^T$$

$$(9.17)$$

We have

$$\frac{\partial D}{\partial x_j} = 0$$

$$\text{for} \quad y^T A v_j + v_j^T A^T y = v_j^T A^T Ax + x^T A^T A v_j$$

But because we have the following scalar property.

$$P = a^T b = (a^T b)^T = b^T a \qquad (9.18)$$

The previous expression can be written as follows.

$$v_j^T (A^T y) = \left[v_j^T (A^T y) \right]^T = (A^T y)^T v_j = y^T A v_j$$

$$v_j^T (A^T A \beta) = \left[v_j^T (A^T Ax) \right]^T = (A^T Ax)^T v_j = x_j^T (A^T A)^T v_j = x_j^T (A^T A) v_j \qquad (9.19)$$

We have

$$y^T A v_j + v_j^T A^T y = 2 v_j^T A^T y$$

$$v_j^T A^T Ax + x^T A^T A v_j = 2 v_j^T A^T Ax$$

$$\text{and}$$

$$2 v_j^T A^T y = 2 v_j^T A^T A \beta$$

whose solution is

$$A^T y = A^T A x$$
$$x = (A^T A)^{-1} A^T y$$

And

$$A x = Q y = A(A^T A)^{-1} A^T y \qquad (9.20)$$

By the pseudo-inverse property we can find among all possible vectors there exists the given invariant vector and the minimal distance between the given or wanted output y as sources and the internal value with given property.

9.2   Geometric Image of the Pseudo-Inverse

We remark that the operator $Q = \dot{A}(A^T A)^{-1} A^T$ is a projection operator for which (Fig. 9.5)

$$Q^2 = A(A^T A)^{-1} A^T A(A^T A)^{-1} A^T = A(A^T A)^{-1} A^T = Q \qquad (9.21)$$

Another possible solution for (9.7) is

$$Z = (B^T A)^{-1} B^T \qquad (9.22)$$

And

$$(B^T A)^{-1} B^T A = Identity$$

$$x = Z A x = (B^T A)^{-1} B^T (A x)$$
$$Q y = A(B^T A)^{-1} B^T y \qquad (9.23)$$

where Qy is an oblique projection operator.

Fig. 9.5 Geometric image of the projection operator

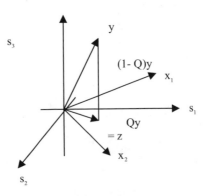

Given the operator A, the pseudo inverse is (9.24)

$$(B^T A)^{-1} B^T \tag{9.24}$$

In fact we have

$$(B^T A)^{-1} B^T [A] = Identity \tag{9.25}$$

Now for the previous pseudo inverse we have (9.26)

$$A(B^T A)^{-1} B^T [AX] = AX \tag{9.26}$$

Or better

$$A(B^T A)^{-1} B^T y = Qy$$

and

$$A(y)(B^T A(y))^{-1} B^T y = Q(y) \, y$$

That can be represented by the input and output system (conceptual filter). So we have a non-linear projection operator where the internal operator is function of the external input value for which we have (9.27).

$$A(y)(B^T A(y))^{-1} B^T y = Q(y)y = z$$
$$A(z)(B^T A(z))^{-1} B^T A(y)(B^T A(y))^{-1} B^T y = Q(z)Q(y)y \tag{9.27}$$
$$A(z)(B^T A(z))^{-1} B^T z$$

Example 9.1 For the two dimensional space we have

$$Y = \begin{bmatrix} \frac{1}{3} \\ \frac{2}{3} \end{bmatrix}, \quad B = \begin{bmatrix} \frac{\sqrt{2}}{2} \\ \frac{\sqrt{2}}{2} \end{bmatrix}, \quad A = \begin{bmatrix} \frac{\sqrt{3}}{2} \\ \frac{1}{2} \end{bmatrix}$$

The geometric representation of the projection operators is shown in Fig. 9.6. In Fig. 9.6, orthogonal projection is

$$A \left(A^T A \right)^{-1} A^T Y = Q_1 Y,$$

Oblique projection is

$$A \left(B^T A \right)^{-1} B^T Y = Q_2 Y$$

The chip figure is shown in Fig. 9.7.

Fig. 9.6 Geometric image of
orthogonal projection and
oblique projection

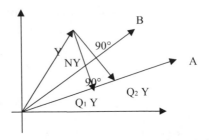

Another possible solution for (9.7) is

$$B = (A^T ZA)^{-1} A^T, \quad y = ZAx, \quad BZA = (A^T ZA)^{-1} A^T ZA = Identity \qquad (9.28)$$

And

$$ZAx = ZA(A^T ZA)^{-1} A^T (ZAx), \quad Qy = ZA(A^T ZA)^{-1} A^T y \qquad (9.29)$$

Morphogenetic-Chip is shown in Fig. 9.8.

So the first projection operator is orthogonal to the $N = I - Q_1$ and the second $N = I - Q_2$ is orthogonal to the orthogonal projection Q_y into B. The orthogonal space to A is

$$N1 = (I - Q_1) \qquad (9.30)$$

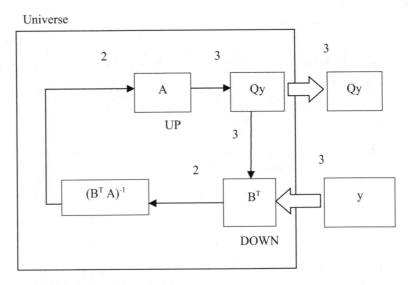

Fig. 9.7 The chip figure of Example 9.1

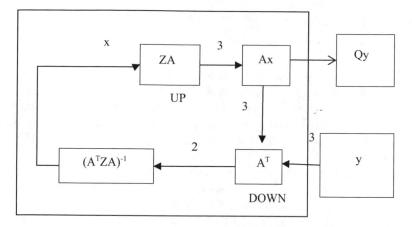

Fig. 9.8 Morphogenetic-Chip

In fact we have

$$A^T\left(I - A(A^TA)^{-1}A^T\right) = A^T - A^TA(A^TA)^{-1}A^T = 0 \qquad (9.31)$$

N1 is orthogonal to A and is the null space whose dimension is N − n, where n is the dimension of the column space A and N is the dimension of the multidimensional space.

Example 9.2 For

$$A = \begin{bmatrix} \frac{\sqrt{2}}{2} \\ \frac{\sqrt{2}}{2} \end{bmatrix}, \quad N = I - A(A^TA)^{-1}A = \begin{bmatrix} \frac{1}{2} & -\frac{1}{2} \\ -\frac{1}{2} & \frac{1}{2} \end{bmatrix}$$

In a graphic way we have Fig. 9.9.

Fig. 9.9 The two orthogonal vectors of A

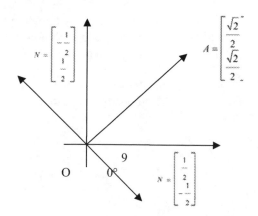

9.3   Simple Genetic Algorithm Subject to Constraint
[22, 23, 25, 26, 35]

The GA has been used in many machine-learning contexts, such as evolving classification rules, evolving neural networks, classifier systems, and automatic programming. In may of these cases there is no closed-form fitness function; the evaluation of each individual (or collection of individuals) is obtained by running it on the particular task being learned. The GA is considered to be successful if an individual, or collection of individuals has satisfactorily learned the given task. The lack of a closed-form fitness function in these problems makes it difficult to study GA performance rigorously. Thus, much of the existing GA theory has been developed in the context of optimization in which the GA is considered to be successful if it discovers a single bit string that represents a value yielding an optimum (or near optimum) of the given function. In this part we create close form or invariant for fitness functions or for other functions. The first closed form or constrain is the normalization property for which we have (9.32).

$$p_1 + p_2 + \cdots + p_n = 1 \qquad (9.32)$$

Other can be average values by (9.33).

$$\langle f(x) \rangle = \sum_i p_i f(x_i) = k \qquad (9.33)$$

Or in a more general form (9.34)

$$F(p_1, \ldots, p_n | q_1, \ldots, q_m) = k \qquad (9.34)$$

9.3.1   Genetic Selection Algorithm in Two Dimensional State Subject to Normalize Constraint

Based on the morphogenetic transformation in mathematical appendix, we can get the selection of morphogenetic evolution in the following way. Given two states 1 and 0 so we have the state space.

S = {1, 0}

Now we know that at any state we can associate one probability for which we have the invariant property or constraint.

$$f(p_1, p_2) = p_1 + p_2 = 1 \tag{9.35}$$

So for the invariant property we have (9.36)

$$\frac{df}{dt} = \frac{dp_1}{dt}\frac{\partial f}{\partial p_1} + \frac{dp_2}{dt}\frac{\partial f}{\partial p_2} = 0 \tag{9.36}$$

In the special case of Eq. (9.36) we have (9.37).

$$Df = \begin{bmatrix} \frac{dp_1}{dt} \\ \frac{dp_2}{dt} \end{bmatrix}^T \begin{bmatrix} \frac{\partial f}{\partial p_1} \\ \frac{\partial f}{\partial p_2} \end{bmatrix} = 0 \tag{9.37}$$

So we have (9.38).

$$Df = \begin{bmatrix} \frac{dp_1}{dt} \\ \frac{dp_2}{dt} \end{bmatrix}^T \begin{bmatrix} 1 \\ 1 \end{bmatrix} = 0 \tag{9.38}$$

For the orthogonal vectors we have (9.39).

$$\begin{bmatrix} \frac{dp_1}{dt} \\ \frac{dp_2}{dt} \end{bmatrix} = \begin{bmatrix} 1 & 0 \\ 0 & 1 \end{bmatrix} - \begin{bmatrix} 1 \\ 1 \end{bmatrix} \left(\begin{bmatrix} 1 \\ 1 \end{bmatrix}^T \begin{bmatrix} 1 \\ 1 \end{bmatrix} \right)^{-1} \begin{bmatrix} 1 \\ 1 \end{bmatrix}^T = \frac{1}{2} \begin{bmatrix} 1 & -1 \\ -1 & 1 \end{bmatrix} \tag{9.39}$$

The variation of the probability has two possible directions, one is down and the other is up on the line where the invariant form is (9.40). So we have two systems of differential equation for which (9.40) is invariant. The two systems of differential equations are

$$\begin{cases} \frac{dp_1}{dt} = \frac{1}{2} \\ \frac{dp_2}{dt} = -\frac{1}{2} \end{cases}, \quad \begin{cases} \frac{dp_1}{dt} = -\frac{1}{2} \\ \frac{dp_2}{dt} = \frac{1}{2} \end{cases} \tag{9.40}$$

Whose solutions are (9.41).

$$\begin{cases} p_1 = \frac{1}{2}t + c_1 \\ p_2 = -\frac{1}{2}t + c_2 \end{cases}, \quad \begin{cases} p_1 = -\frac{1}{2}t + c_3 \\ p_2 = \frac{1}{2}t + c_4 \end{cases} \tag{9.41}$$

In Fig. 9.10, we can see the variation of the probability down from p(t) to Tp(t).
We have the simplex space given of the line where (9.40) is true. Now we begin with the initial vector p(t) and we move from p(t) to the final vector.

$$Tp = \begin{bmatrix} F_1 & 0 \\ 0 & F_2 \end{bmatrix} \begin{bmatrix} p_1 \\ p_2 \end{bmatrix}$$

Fig. 9.10 The relation
between p(t) and p(t + 1)

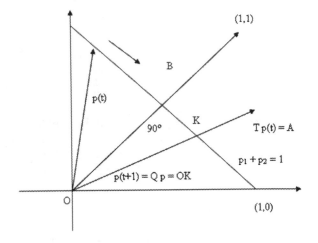

where T is the Michael Vose transformation for selection in the simple genetic algorithm. The selection vector or fitness vector is $\begin{bmatrix} F_1 \\ F_2 \end{bmatrix}$. The operator that moves from p(t) to Tp(t) in the line of probability for which (9.40) is true, is an oblique projection given by the operator.

$$A(p(t))(B^T A(p(t)))^{-1} B^T y(t) = Q(p(t))\, p(t) = p(t+1) \qquad (9.41)$$

For the very simple example we have

$$p = \begin{bmatrix} p_1 \\ p_2 \end{bmatrix}, \quad A(p) = \begin{bmatrix} F_1 & 0 \\ 0 & F_2 \end{bmatrix} \begin{bmatrix} p_1 \\ p_2 \end{bmatrix}, \quad B = \begin{bmatrix} 1 \\ 1 \end{bmatrix}$$

The recursion process with constraint is given by the expression (9.42).

$$\begin{bmatrix} p_1(t+1) \\ p_2(t+1) \end{bmatrix} = \begin{bmatrix} F_1 & 0 \\ 0 & F_2 \end{bmatrix} \begin{bmatrix} p_1(t) \\ p_2(t) \end{bmatrix} \left(\begin{bmatrix} 1 \\ 1 \end{bmatrix}^T \begin{bmatrix} F_1 & 0 \\ 0 & F_2 \end{bmatrix} \begin{bmatrix} p_1(t) \\ p_2(t) \end{bmatrix} \right)^{-1} \begin{bmatrix} 1 \\ 1 \end{bmatrix}^T \begin{bmatrix} p_1(t) \\ p_2(t) \end{bmatrix}$$

$$= \begin{bmatrix} \frac{F_1 p_1(t)}{F_1 p_1(t) + F_2 p_2(t)} (p_1(t) + p_2(t)) \\ \frac{F_2 p_2(t)}{F_1 p_1(t) + F_2 p_2(t)} (p_1(t) + p_2(t)) \end{bmatrix}$$

$$(9.42)$$

The expression is the same of the M.D. Vose and J.E. Rowe with the difference that we use the expression of the oblique projection where the transformation of the probability is orthogonal to the vector B for which the probability constraint of the normalization is always true. So the transformation is always subject to normalization constraint in any moment. In fact we can see that for any step we have that the invariant form or constraint is always true.

$$p_1 + p_2 = 1, \quad and \quad p(t+1) = \begin{bmatrix} \frac{F_1 p_1(t)}{F_1 p_1(t) + F_2 p_2(t)} \\ \frac{F_2 p_2(t)}{F_1 p_1(t) + F_2 p_2(t)} \end{bmatrix},$$

Where

$$\frac{F_1 p_1(t)}{F_1 p_1(t) + F_2 p_2(t)} + \frac{F_2 p_2(t)}{F_1 p_1(t) + F_2 p_2(t)} = 1$$

$B = \begin{bmatrix} 1 \\ 1 \end{bmatrix}$, so for the property of the oblique projection any projection or movement is orthogonal to the vector B as we can see in the differential equations for the probability constraint in Eq. (9.43).

$$\begin{bmatrix} F_1 \\ F_2 \end{bmatrix} = \begin{bmatrix} \frac{\partial f}{\partial p_1} \\ \frac{\partial f}{\partial p_2} \end{bmatrix} = \begin{bmatrix} 1 \\ 1 \end{bmatrix} \tag{9.43}$$

We have

$$p = \begin{bmatrix} p_1 \\ p_2 \end{bmatrix}, A(p) = \begin{bmatrix} 1 & 0 \\ 0 & 1 \end{bmatrix} \begin{bmatrix} p_1 \\ p_2 \end{bmatrix} = \begin{bmatrix} p_1 \\ p_2 \end{bmatrix} = p, B = \begin{bmatrix} 1 \\ 1 \end{bmatrix}$$

So the projection of p into the vector A(p) is again p because p belongs to the vector A(p). So

$$B = \begin{bmatrix} \frac{\partial f}{\partial p_1} \\ \frac{\partial f}{\partial p_2} \end{bmatrix} = \begin{bmatrix} 1 \\ 1 \end{bmatrix}$$

defines the invariance property as the orthogonality between the velocity vector of p and B. The vector A(p) is the aim of the oblique projection that starts from p. When A(p) = p the initial and final value of the projection are the same so any initial value of the probability is a fixed value in the recursion process. So we have (9.44).

$$Qp(t) = \begin{bmatrix} \frac{F_1 p_1(t)}{F_1 p_1(t) + F_2 p_2(t)} \\ \frac{F_2 p_2(t)}{F_1 p_1(t) + F_2 p_2(t)} \end{bmatrix} = \begin{bmatrix} p_1 \\ p_2 \end{bmatrix} = p(t) \tag{9.44}$$

where we have the two fixed points

$$\begin{bmatrix} p_1 \\ p_2 \end{bmatrix} = \begin{bmatrix} 1 \\ 0 \end{bmatrix}, \quad \begin{bmatrix} p_1 \\ p_2 \end{bmatrix} = \begin{bmatrix} 0 \\ 1 \end{bmatrix}$$

Now there are many possible ways which are orthogonal to B. In fact we have two possible ways to be orthogonal given by the two vectors.

$$\begin{bmatrix} \frac{dp_1}{dt} \\ \frac{dp_2}{dt} \end{bmatrix} = \frac{1}{2}\begin{bmatrix} 1 \\ -1 \end{bmatrix}, \qquad \begin{bmatrix} \frac{dp_1}{dt} \\ \frac{dp_2}{dt} \end{bmatrix} = \frac{1}{2}\begin{bmatrix} -1 \\ 1 \end{bmatrix}$$

In a graph way we have Fig. 9.11.

When $F_1 > F_2$ we have that $\begin{bmatrix} p_1 \\ p_2 \end{bmatrix} = \begin{bmatrix} 1 \\ 0 \end{bmatrix}$ is a stable point and $\begin{bmatrix} p_1 \\ p_2 \end{bmatrix} = \begin{bmatrix} 0 \\ 1 \end{bmatrix}$ is an unstable point as we can see by the expression (9.45).

$$\begin{bmatrix} \frac{F_1 p_1(t)}{F_1 p_1(t) + F_2 p_2(t)} \\ \frac{F_2 p_2(t)}{F_1 p_1(t) + F_2 p_2(t)} \end{bmatrix} - \begin{bmatrix} p_1 \\ p_2 \end{bmatrix} = \begin{bmatrix} \frac{F_1 p_1(t)}{F_1 p_1(t) + F_2 p_2(t)} - p_1 \\ \frac{F_2 p_2(t)}{F_1 p_1(t) + F_2 p_2(t)} - p_2 \end{bmatrix} = \begin{bmatrix} \left(\frac{F_1}{F_1 p_1(t) + F_2 p_2(t)} - 1\right)p_1 \\ \left(\frac{F_2}{F_1 p_1(t) + F_2 p_2(t)} - 1\right)p_2 \end{bmatrix}$$

$$(9.45)$$

In a graphic way we have Fig. 9.12.

The projection of $\begin{bmatrix} F_1 & 0 \\ 0 & F_2 \end{bmatrix}\begin{bmatrix} 1 \\ 1 \end{bmatrix} = \begin{bmatrix} F_1 \\ F_2 \end{bmatrix} = F$ into the orthogonal vector $\begin{bmatrix} 1 \\ -1 \end{bmatrix}$ is

$$QF = \begin{bmatrix} 1 \\ -1 \end{bmatrix}\left(\begin{bmatrix} 1 \\ -1 \end{bmatrix}^T\begin{bmatrix} 1 \\ -1 \end{bmatrix}\right)^{-1}\begin{bmatrix} 1 \\ -1 \end{bmatrix}^T\begin{bmatrix} F_1 \\ F_2 \end{bmatrix} = \frac{1}{2}\begin{bmatrix} F_1 - F_2 \\ -(F_1 - F_2) \end{bmatrix}$$

Because $F_1 > F_2$ the projection has the same direction of the movement of the evolution. With the previous projection we can create another constraint in the

Fig. 9.11 Two possible ways to be orthogonal

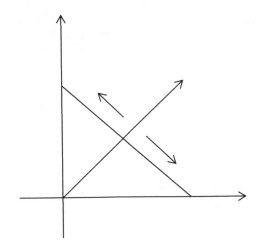

Fig. 9.12 Evolution process
to the best feet values F

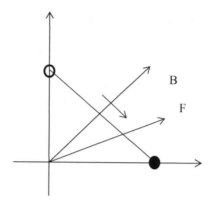

initial condition in a way to make a right direction to the evolution to obtain the
maximum of the probability for the maximum value of the fitness function in a
rapid way. So we choose the initial condition with the given rule.

$$p(0) = \frac{1}{2}\begin{bmatrix} 1 \\ 1 \end{bmatrix} + \frac{1}{2}\begin{bmatrix} F_1 - F_2 \\ -(F_1 - F_2) \end{bmatrix}\Delta t$$

Figure 9.13 is the image of the initial condition.

When $F_1 < F_2$ we have that $\begin{bmatrix} p_1 \\ p_2 \end{bmatrix} = \begin{bmatrix} 1 \\ 0 \end{bmatrix}$ is an unstable point and $\begin{bmatrix} p_1 \\ p_2 \end{bmatrix} = \begin{bmatrix} 0 \\ 1 \end{bmatrix}$
is a stable point (Fig. 9.14).

Fig. 9.13 The initial
condition

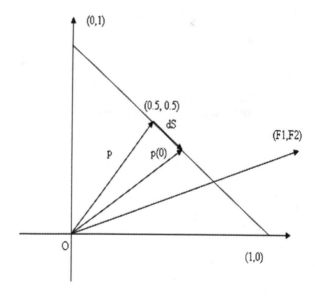

Fig. 9.14 Evolution back to
the feet values F

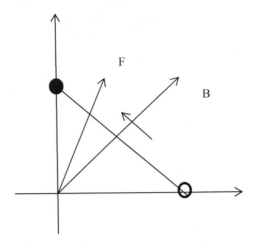

We introduce the same initial condition by projection operator.

$$p(0) = \frac{1}{2}\begin{bmatrix} 1 \\ 1 \end{bmatrix} + \frac{1}{2}\begin{bmatrix} F_1 - F_2 \\ -(F_1 - F_2) \end{bmatrix}\Delta t$$

Because now $F_1 < F_2$ the direction of the initial value of the probability is the reverse to the previous case so the initial probability constraint controls the direction of the evolution in a way to be directed to the stable point where we have the maximum value of the fitness function.

The stability or instability of the fixed points is the function of the position of the fitness function. For example, given

$$p = \begin{bmatrix} \frac{1}{4} \\ \frac{3}{4} \end{bmatrix}, \quad A = \begin{bmatrix} 4 & 0 \\ 0 & 1 \end{bmatrix}\begin{bmatrix} \frac{1}{4} \\ \frac{3}{4} \end{bmatrix}$$

We have the process Fig. 9.15.
For initial condition directed to the stable point we have

$$p = \begin{bmatrix} \frac{3}{4} \\ \frac{1}{4} \end{bmatrix}, \quad A = \begin{bmatrix} 4 & 0 \\ 0 & 1 \end{bmatrix}\begin{bmatrix} \frac{3}{4} \\ \frac{1}{4} \end{bmatrix}$$

We have the evolution (Fig. 9.16).

Fig. 9.15 Probability
evolution to the best feet
value in *red*
(color figure online)

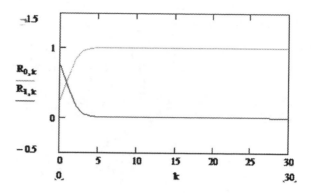

Fig. 9.16 Evolution of the
probability when we change
the initial probability values

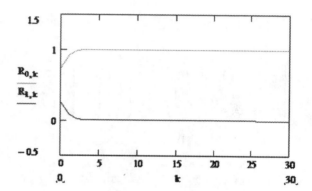

9.3.2   Three Dimension Selection Evolution Subject to the Normalize Constraint

Example 9.3 For n = 3 we have the constraint.

$$p_1 + p_2 + p_3 = 1$$

And the evolution subject to constraint is shown in (9.46).

$$\begin{bmatrix} F_1 & 0 & 0 \\ 0 & F_2 & 0 \\ 0 & 0 & F_3 \end{bmatrix} \begin{bmatrix} 1 \\ 1 \\ 1 \end{bmatrix} \left(\begin{bmatrix} 1 \\ 1 \\ 1 \end{bmatrix}^T \begin{bmatrix} F_1 & 0 & 0 \\ 0 & F_2 & 0 \\ 0 & 0 & F_3 \end{bmatrix} \begin{bmatrix} 1 \\ 1 \\ 1 \end{bmatrix} \right)^{-1} \begin{bmatrix} 1 \\ 1 \\ 1 \end{bmatrix}^T \begin{bmatrix} p_1(t) \\ p_2(t) \\ p_3(t) \end{bmatrix}$$

$$= \begin{bmatrix} \dfrac{F_1 p_1(t)}{F_1 p_1(t) + F_1 p_1(t) + F_1 p_1(t)} \\ \dfrac{F_2 p_2(t)}{F_1 p_1(t) + F_1 p_1(t) + F_1 p_1(t)} \\ \dfrac{F_3 p_3(t)}{F_1 p_1(t) + F_1 p_1(t) + F_1 p_1(t)} \end{bmatrix} = \begin{bmatrix} p_1(t+1) \\ p_2(t+1) \\ p_2(t+1) \end{bmatrix}$$

$$(9.46)$$

where the fixed points are

$$p_1 = \begin{bmatrix} 1 \\ 0 \\ 0 \end{bmatrix}, \quad p_2 = \begin{bmatrix} 0 \\ 1 \\ 0 \end{bmatrix}, \quad p_3 = \begin{bmatrix} 0 \\ 0 \\ 1 \end{bmatrix}$$

Given the gradient vector

$$f = p_1 + p_2 + p_3, \quad \nabla f = \begin{bmatrix} \frac{\partial f}{\partial p_1} \\ \frac{\partial f}{\partial p_2} \\ \frac{\partial f}{\partial p_3} \end{bmatrix} = \begin{bmatrix} 1 \\ 1 \\ 1 \end{bmatrix}.$$

The basis orthogonal vectors to the gradient are obtained by the expression (9.47).

$$\begin{bmatrix} \frac{dp_1}{dt} \\ \frac{dp_2}{dt} \\ \frac{dp_3}{dt} \end{bmatrix} = \begin{bmatrix} 1 & 0 & 0 \\ 0 & 1 & 0 \\ 0 & 0 & 1 \end{bmatrix} - \begin{bmatrix} 1 \\ 1 \\ 1 \end{bmatrix} \left(\begin{bmatrix} 1 \\ 1 \\ 1 \end{bmatrix}^T \begin{bmatrix} 1 \\ 1 \\ 1 \end{bmatrix} \right)^{-1} \begin{bmatrix} 1 \\ 1 \\ 1 \end{bmatrix}^T = \frac{1}{3} \begin{bmatrix} 2 & -1 & -1 \\ -1 & 2 & -1 \\ -1 & -1 & 2 \end{bmatrix}$$

(9.47)

It can be represented by Fig. 9.17.

The three column vectors are not independent. Only two vectors are independent. So we have

$$H = \begin{bmatrix} 2 & -1 \\ -1 & 2 \\ -1 & -1 \end{bmatrix} \frac{1}{3}.$$

It can be given by Fig. 9.18.

Fig. 9.17 Three possible evolution process of the probability for three possible best feet values

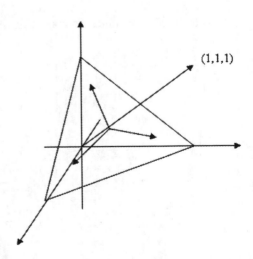

(1,1,1)

Fig. 9.18 Two independent
evolution process for the
probability

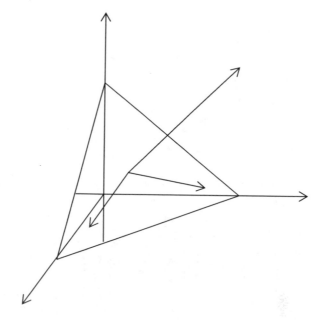

Given the transformation

$$\begin{bmatrix} F_1 & 0 & 0 \\ 0 & F_2 & 0 \\ 0 & 0 & F_3 \end{bmatrix} \begin{bmatrix} 1 \\ 1 \\ 1 \end{bmatrix} = \begin{bmatrix} F_1 \\ F_2 \\ F_3 \end{bmatrix}$$

We have the initial direction constraint.

$$p(0) = \begin{bmatrix} \frac{1}{3} \\ \frac{1}{3} \\ \frac{1}{3} \end{bmatrix} + \Delta p = \begin{bmatrix} \frac{1}{3} \\ \frac{1}{3} \\ \frac{1}{3} \end{bmatrix} + \begin{bmatrix} \frac{2}{3} & -\frac{1}{3} \\ -\frac{1}{3} & \frac{2}{3} \\ -\frac{1}{3} & -\frac{1}{3} \end{bmatrix} \left(\begin{bmatrix} \frac{2}{3} & -\frac{1}{3} \\ -\frac{1}{3} & \frac{2}{3} \\ -\frac{1}{3} & -\frac{1}{3} \end{bmatrix}^T \begin{bmatrix} \frac{2}{3} & -\frac{1}{3} \\ -\frac{1}{3} & \frac{2}{3} \\ -\frac{1}{3} & -\frac{1}{3} \end{bmatrix} \right)^{-1} \begin{bmatrix} \frac{2}{3} & -\frac{1}{3} \\ -\frac{1}{3} & \frac{2}{3} \\ -\frac{1}{3} & -\frac{1}{3} \end{bmatrix}^T$$

$$\begin{bmatrix} F_1 \\ F_2 \\ F_3 \end{bmatrix} \Delta t = \begin{bmatrix} \frac{1}{3} \\ \frac{1}{3} \\ \frac{1}{3} \end{bmatrix} + \begin{bmatrix} \frac{2}{3}F_1 - \frac{1}{3}F_2 - \frac{1}{3}F_3 \\ -\frac{1}{3}F_1 + \frac{2}{3}F_2 - \frac{1}{3}F_3 \\ -\frac{1}{3}F_1 - \frac{1}{3}F_2 + \frac{2}{3}F_3 \end{bmatrix} \Delta t$$

$$(9.48)$$

For the eight populations (000), (001), (010), (011), (100), (101), (110), (111),
the initial probabilities are

$$
\begin{pmatrix} s1_0 \\ s2_0 \\ s3_0 \\ s4_0 \\ s5_0 \\ s6_0 \\ s7_0 \\ s8_0 \end{pmatrix} := \begin{pmatrix} \frac{1}{8} \\ \frac{1}{8} \\ \frac{1}{8} \\ \frac{1}{8} \\ \frac{1}{8} \\ \frac{1}{8} \\ \frac{1}{8} \\ \frac{1}{8} \end{pmatrix} + \begin{pmatrix} \frac{7}{8}\cdot F1 - \frac{1}{8}\cdot F2 - \frac{1}{8}\cdot F3 - \frac{1}{8}\cdot F4 - \frac{1}{8}\cdot F5 - \frac{1}{8}\cdot F6 - \frac{1}{8}\cdot F7 - \frac{1}{8}\cdot F8 \\ \frac{-1}{8}\cdot F1 + \frac{7}{8}\cdot F2 - \frac{1}{8}\cdot F3 - \frac{1}{8}\cdot F4 - \frac{1}{8}\cdot F5 - \frac{1}{8}\cdot F6 - \frac{1}{8}\cdot F7 - \frac{1}{8}\cdot F8 \\ \frac{-1}{8}\cdot F1 - \frac{1}{8}\cdot F2 + \frac{7}{8}\cdot F3 - \frac{1}{8}\cdot F4 - \frac{1}{8}\cdot F5 - \frac{1}{8}\cdot F6 - \frac{1}{8}\cdot F7 - \frac{1}{8}\cdot F8 \\ \frac{-1}{8}\cdot F1 - \frac{1}{8}\cdot F2 - \frac{1}{8}\cdot F3 + \frac{7}{8}\cdot F4 - \frac{1}{8}\cdot F5 - \frac{1}{8}\cdot F6 - \frac{1}{8}\cdot F7 - \frac{1}{8}\cdot F8 \\ \frac{-1}{8}\cdot F1 - \frac{1}{8}\cdot F2 - \frac{1}{8}\cdot F3 - \frac{1}{8}\cdot F4 + \frac{7}{8}\cdot F5 - \frac{1}{8}\cdot F6 - \frac{1}{8}\cdot F7 - \frac{1}{8}\cdot F8 \\ \frac{-1}{8}\cdot F1 - \frac{1}{8}\cdot F2 - \frac{1}{8}\cdot F3 - \frac{1}{8}\cdot F4 - \frac{1}{8}\cdot F5 - \frac{7}{8}\cdot F6 + \frac{1}{8}\cdot F7 - \frac{1}{8}\cdot F8 \\ \frac{-1}{8}\cdot F1 - \frac{1}{8}\cdot F2 - \frac{1}{8}\cdot F3 - \frac{1}{8}\cdot F4 - \frac{1}{8}\cdot F5 - \frac{1}{8}\cdot F6 + \frac{7}{8}\cdot F7 - \frac{1}{8}\cdot F8 \\ \frac{-1}{8}\cdot F1 - \frac{1}{8}\cdot F2 - \frac{1}{8}\cdot F3 - \frac{1}{8}\cdot F4 - \frac{1}{8}\cdot F5 - \frac{1}{8}\cdot F6 - \frac{1}{8}\cdot F7 + \frac{7}{8}\cdot F8 \end{pmatrix} \cdot da
$$

The morphogenetic eight behaviors of the probabilities in the ordinates for the eight populations is shown in Fig. 9.19.

Where only the last population with the maximum fitness assumes the maximum value of the probability in 20 generations. The fitness vector is

$$
F = \begin{bmatrix} F_1 \\ F_2 \\ F_3 \\ F_4 \\ F_5 \\ F_6 \\ F_7 \\ F_8 \end{bmatrix} = \begin{bmatrix} 3 \\ 2 \\ 2 \\ 1 \\ 2 \\ 1 \\ 1 \\ 4 \end{bmatrix}
$$

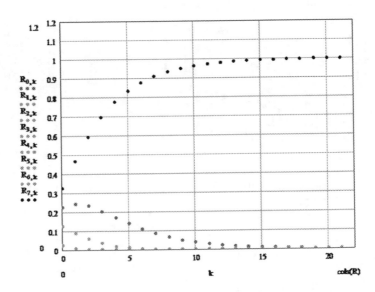

Fig. 9.19 Evolution of the probability for 8 fitness values

9.4   Selection, Mutation and Crossover Evolution Subject to the Normalize Constraint and Initial Probability Constraint

Given the mutation transformations (9.49) (M.D. Vose and J.E. Rowe)

$$
\begin{bmatrix}
(1-\mu)^2 & (1-\mu)\mu & (1-\mu)\mu & \mu^2 \\
(1-\mu)\mu & (1-\mu)^2 & \mu^2 & (1-\mu)\mu \\
(1-\mu)\mu & \mu^2 & (1-\mu)^2 & (1-\mu)\mu \\
\mu^2 & (1-\mu)\mu & (1-\mu)\mu & (1-\mu)^2
\end{bmatrix} = U \qquad (9.49)
$$

where

$$
\begin{bmatrix}
F_1 & 0 & 0 & 0 \\
0 & F_2 & 0 & 0 \\
0 & 0 & F_3 & 0 \\
0 & 0 & 0 & F_4
\end{bmatrix} = F
$$

And the composition is (9.50).

$$
\begin{aligned}
UF &=
\begin{bmatrix}
(1-\mu)^2 & (1-\mu)\mu & (1-\mu)\mu & \mu^2 \\
(1-\mu)\mu & (1-\mu)^2 & \mu^2 & (1-\mu)\mu \\
(1-\mu)\mu & \mu^2 & (1-\mu)^2 & (1-\mu)\mu \\
\mu^2 & (1-\mu)\mu & (1-\mu)\mu & (1-\mu)^2
\end{bmatrix}
\begin{bmatrix}
F_1 & 0 & 0 & 0 \\
0 & F_2 & 0 & 0 \\
0 & 0 & F_3 & 0 \\
0 & 0 & 0 & F_4
\end{bmatrix} \\
&=
\begin{bmatrix}
F_1(1-\mu)^2 & F_2(1-\mu)\mu & F_3(1-\mu)\mu & F_4\mu^2 \\
F_1(1-\mu)\mu & F_2(1-\mu)^2 & F_3\mu^2 & F_4(1-\mu)\mu \\
F_1(1-\mu)\mu & F_2\mu^2 & F_3(1-\mu)^2 & F_4(1-\mu)\mu \\
F_1\mu^2 & F_2(1-\mu)\mu & F_3(1-\mu)\mu & F_4(1-\mu)^2
\end{bmatrix}
\end{aligned}
$$

$$(9.50)$$

So the evolution with the normalization constraint is (9.51).

$$
f = p_1 + p_2 + p_3 + p_4 = 1
$$

$$
\begin{bmatrix} p_1(t+1) \\ p_2(t+1) \\ p_3(t+1) \\ p_4(t+1) \end{bmatrix} = \begin{bmatrix} F_1(1-\mu)^2 & F_2(1-\mu)\mu & F_3(1-\mu)\mu & F_4\mu^2 \\ F_1(1-\mu)\mu & F_2(1-\mu)^2 & F_3\mu^2 & F_4(1-\mu)\mu \\ F_1(1-\mu)\mu & F_2\mu^2 & F_3(1-\mu)^2 & F_4(1-\mu)\mu \\ F_1\mu^2 & F_2(1-\mu)\mu & F_3(1-\mu)\mu & F_4(1-\mu)^2 \end{bmatrix} \begin{bmatrix} p_1(t) \\ p_2(t) \\ p_3(t) \\ p_4(t) \end{bmatrix}
$$

$$
\left(\begin{bmatrix} 1 \\ 1 \\ 1 \\ 1 \end{bmatrix}^T \begin{bmatrix} F_1(1-\mu)^2 & F_2(1-\mu)\mu & F_3(1-\mu)\mu & F_4\mu^2 \\ F_1(1-\mu)\mu & F_2(1-\mu)^2 & F_3\mu^2 & F_4(1-\mu)\mu \\ F_1(1-\mu)\mu & F_2\mu^2 & F_3(1-\mu)^2 & F_4(1-\mu)\mu \\ F_1\mu^2 & F_2(1-\mu)\mu & F_3(1-\mu)\mu & F_4(1-\mu)^2 \end{bmatrix} \begin{bmatrix} p_1(t) \\ p_2(t) \\ p_3(t) \\ p_4(t) \end{bmatrix} \right)^{-1} \begin{bmatrix} 1 \\ 1 \\ 1 \\ 1 \end{bmatrix}^T \begin{bmatrix} p_1(t) \\ p_2(t) \\ p_3(t) \\ p_4(t) \end{bmatrix}
$$

$$(9.51)$$

And the initial condition is given by the expression (9.52).

$$
\begin{bmatrix} \frac{dp_1}{dt} \\ \frac{dp_2}{dt} \\ \frac{dp_3}{dt} \\ \frac{dp_4}{dt} \end{bmatrix} = \begin{bmatrix} 1 & 0 & 0 & 0 \\ 0 & 1 & 0 & 0 \\ 0 & 0 & 1 & 0 \\ 0 & 0 & 0 & 1 \end{bmatrix} - \begin{bmatrix} 1 \\ 1 \\ 1 \\ 1 \end{bmatrix} \left(\begin{bmatrix} 1 \\ 1 \\ 1 \\ 1 \end{bmatrix}^T \begin{bmatrix} 1 \\ 1 \\ 1 \\ 1 \end{bmatrix} \right)^{-1} \begin{bmatrix} 1 \\ 1 \\ 1 \\ 1 \end{bmatrix}^T
$$

$$
= \frac{1}{4} \begin{bmatrix} 3 & -1 & -1 & -1 \\ -1 & 3 & -1 & -1 \\ -1 & -1 & 3 & -1 \\ -1 & -1 & -1 & 3 \end{bmatrix}
$$

$$(9.52)$$

Given the transformation (9.53)

$$
\begin{bmatrix} F_1(1-\mu)^2 & F_2(1-\mu)\mu & F_3(1-\mu)\mu & F_4\mu^2 \\ F_1(1-\mu)\mu & F_2(1-\mu)^2 & F_3\mu^2 & F_4(1-\mu)\mu \\ F_1(1-\mu)\mu & F_2\mu^2 & F_3(1-\mu)^2 & F_4(1-\mu)\mu \\ F_1\mu^2 & F_2(1-\mu)\mu & F_3(1-\mu)\mu & F_4(1-\mu)^2 \end{bmatrix} \begin{bmatrix} 1 \\ 1 \\ 1 \\ 1 \end{bmatrix}
$$

$$
= \begin{bmatrix} F_1(1-\mu)^2 + F_2(1-\mu)\mu + F_3(1-\mu)\mu + F_4\mu^2 \\ F_1(1-\mu)\mu + F_2(1-\mu)^2 + F_3\mu^2 + F_4(1-\mu)\mu \\ F_1(1-\mu)\mu + F_2\mu^2 + F_3(1-\mu)^2 + F_4(1-\mu)\mu \\ F_1\mu^2 + F_2(1-\mu)\mu + F_3(1-\mu)\mu + F_4(1-\mu)^2 \end{bmatrix}
$$

$$(9.53)$$

We have the projection on the orthogonal independent vectors in a way to establish a direction constraint on the initial value of the probability.

$$p(0) = \begin{bmatrix} 0.25 \\ 0.25 \\ 0.25 \\ 0.25 \end{bmatrix} + \Delta p = \begin{bmatrix} 0.25 \\ 0.25 \\ 0.25 \\ 0.25 \end{bmatrix}$$

$$+ \left(\begin{bmatrix} \frac{3}{4} & -\frac{1}{4} & -\frac{1}{4} \\ -\frac{1}{4} & \frac{3}{4} & -\frac{1}{4} \\ -\frac{1}{4} & -\frac{1}{4} & \frac{3}{4} \\ -\frac{1}{4} & -\frac{1}{4} & -\frac{1}{4} \end{bmatrix} \begin{bmatrix} \frac{3}{4} & -\frac{1}{4} & -\frac{1}{4} \\ -\frac{1}{4} & \frac{3}{4} & -\frac{1}{4} \\ -\frac{1}{4} & -\frac{1}{4} & \frac{3}{4} \\ -\frac{1}{4} & -\frac{1}{4} & -\frac{1}{4} \end{bmatrix}^T \begin{bmatrix} \frac{3}{4} & -\frac{1}{4} & -\frac{1}{4} \\ -\frac{1}{4} & \frac{3}{4} & -\frac{1}{4} \\ -\frac{1}{4} & -\frac{1}{4} & \frac{3}{4} \\ -\frac{1}{4} & -\frac{1}{4} & -\frac{1}{4} \end{bmatrix} \right)^{-1}$$

$$\begin{bmatrix} \frac{3}{4} & -\frac{1}{4} & -\frac{1}{4} \\ -\frac{1}{4} & \frac{3}{4} & -\frac{1}{4} \\ -\frac{1}{4} & -\frac{1}{4} & \frac{3}{4} \\ -\frac{1}{4} & -\frac{1}{4} & -\frac{1}{4} \end{bmatrix}^T \begin{bmatrix} F_1(1-\mu)^2 + F_2(1-\mu)\mu + F_3(1-\mu)\mu + F_4\mu^2 \\ F_1(1-\mu)\mu + F_2(1-\mu)^2 + F_3\mu^2 + F_4(1-\mu)\mu \\ F_1(1-\mu)\mu + F_2\mu^2 + F_3(1-\mu)^2 + F_4(1-\mu)\mu \\ F_1\mu^2 + F_2(1-\mu)\mu + F_3(1-\mu)\mu + F_4(1-\mu)^2 \end{bmatrix} \Delta t$$

For the cross over transformation we have the probability matrix (9.54) for the state (00), (01), (10), (11)

$$M(00) = \begin{bmatrix} 1 & \frac{1}{2} & \frac{1}{2} & \frac{1}{4} \\ \frac{1}{2} & 0 & \frac{1}{4} & 0 \\ \frac{1}{2} & \frac{1}{4} & 0 & 0 \\ \frac{1}{4} & 0 & 0 & 0 \end{bmatrix} = \begin{bmatrix} 1 & \frac{1}{2} \\ \frac{1}{2} & 0 \end{bmatrix} \otimes \begin{bmatrix} 1 & \frac{1}{2} \\ \frac{1}{2} & 0 \end{bmatrix},$$

$$M(01) = \begin{bmatrix} 0 & \frac{1}{2} & 0 & \frac{1}{4} \\ \frac{1}{2} & 1 & \frac{1}{4} & \frac{1}{2} \\ 0 & \frac{1}{4} & 0 & 0 \\ \frac{1}{4} & \frac{1}{2} & 0 & 0 \end{bmatrix} = \begin{bmatrix} 1 & \frac{1}{2} \\ \frac{1}{2} & 0 \end{bmatrix} \otimes \begin{bmatrix} 0 & \frac{1}{2} \\ \frac{1}{2} & 1 \end{bmatrix}$$

$$M(10) = \begin{bmatrix} 0 & 0 & \frac{1}{2} & \frac{1}{4} \\ 0 & 0 & \frac{1}{4} & 0 \\ \frac{1}{2} & \frac{1}{4} & 1 & \frac{1}{2} \\ \frac{1}{4} & 0 & \frac{1}{2} & 0 \end{bmatrix} = \begin{bmatrix} 0 & \frac{1}{2} \\ \frac{1}{2} & 1 \end{bmatrix} \otimes \begin{bmatrix} 1 & \frac{1}{2} \\ \frac{1}{2} & 0 \end{bmatrix},$$

$$M(11) = \begin{bmatrix} 0 & 0 & 0 & \frac{1}{4} \\ 0 & 0 & \frac{1}{4} & \frac{1}{2} \\ 0 & \frac{1}{4} & 0 & \frac{1}{2} \\ \frac{1}{4} & \frac{1}{2} & \frac{1}{2} & 1 \end{bmatrix} = \begin{bmatrix} 0 & \frac{1}{2} \\ \frac{1}{2} & 1 \end{bmatrix} \otimes \begin{bmatrix} 0 & \frac{1}{2} \\ \frac{1}{2} & 1 \end{bmatrix}$$

(9.54)

They are associated to four isomorphic graphs (Fig. 9.20).

Fig. 9.20 Cross over graph

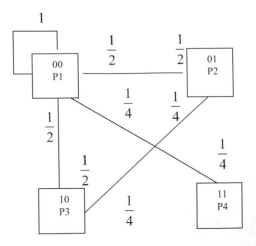

$$M(00) = \begin{bmatrix} 1 & \frac{1}{2} & \frac{1}{2} & \frac{1}{4} \\ \frac{1}{2} & 0 & \frac{1}{4} & 0 \\ \frac{1}{2} & \frac{1}{4} & 0 & 0 \\ \frac{1}{4} & 0 & 0 & 0 \end{bmatrix} = \begin{bmatrix} 1 & \frac{1}{2} \\ \frac{1}{2} & 0 \end{bmatrix} \otimes \begin{bmatrix} 1 & \frac{1}{2} \\ \frac{1}{2} & 0 \end{bmatrix},$$

So for the children probability to be in the state (00) is (9.55).

$$p(00) = p_1 p_1 + \frac{1}{2} p_1 p_2 + \frac{1}{2} p_2 p_1 + \frac{1}{2} p_1 p_3 + \frac{1}{2} p_3 p_1 + \frac{1}{4} p_1 p_4$$
$$+ \frac{1}{4} p_4 p_1 + \frac{1}{4} p_2 p_3 + \frac{1}{4} p_3 p_2 = p_1 p_1 + p_1 p_2 + p_1 p_3 + \frac{1}{2} p_1 p_4 + \frac{1}{2} p_2 p_3$$

$$(9.55)$$

When we repeat the similar process for all the four matrices we have the crossover non-linear transformation of the probability (9.56)

$$\begin{bmatrix} p_{00}(t+1) \\ p_{01}(t+1) \\ p_{10}(t+1) \\ p_{11}(t+1) \end{bmatrix} = \begin{bmatrix} p_1(t+1) \\ p_2(t+1) \\ p_3(t+1) \\ p_4(t+1) \end{bmatrix}$$

$$(9.56)$$

$$= \begin{bmatrix} p_1^2(t) + p_1(t)p_2(t) + p_1(t)p_3(t) + \frac{1}{2}p_1(t)p_4(t) + \frac{1}{2}p_2(t)p_3(t) \\ p_2^2(t) + p_2(t)p_1(t) + p_2(t)p_4(t) + \frac{1}{2}p_2(t)p_3(t) + \frac{1}{2}p_1(t)p_4(t) \\ p_3^2(t) + p_3(t)p_1(t) + p_3(t)p_4(t) + \frac{1}{2}p_3(t)p_2(t) + \frac{1}{2}p_1(t)p_4(t) \\ p_4^2(t) + p_4(t)p_2(t) + p_4(t)p_3(t) + \frac{1}{2}p_4(t)p_1(t) + \frac{1}{2}p_2(t)p_3(t) \end{bmatrix} = C$$

The morphogenetic evolution process is given by the combination of fitness, mutation and crossover transformation in this symbolic expression (9.57).

$$p(t+1) = \frac{f}{\sum_j f_i(t)}, \quad f = CUSp(t) \tag{9.57}$$

$$\begin{bmatrix} p_1^2(t) + p_1(t)p_2(t) + p_1(t)p_3(t) + \frac{1}{2}p_1(t)p_4(t) + \frac{1}{2}p_2(t)p_3(t) \\ p_2^2(t) + p_2(t)p_1(t) + p_2(t)p_4(t) + \frac{1}{2}p_2(t)p_3(t) + \frac{1}{2}p_1(t)p_4(t) \\ p_3^2(t) + p_3(t)p_1(t) + p_3(t)p_4(t) + \frac{1}{2}p_3(t)p_2(t) + \frac{1}{2}p_1(t)p_4(t) \\ p_4^2(t) + p_4(t)p_2(t) + p_4(t)p_3(t) + \frac{1}{2}p_4(t)p_1(t) + \frac{1}{2}p_2(t)p_3(t) \end{bmatrix} = C$$

We have the M.D. Vose and J.E. Rowe complex transformation for which we can give the same genetic evolution process subject to normalized constraint and also the initial probability constraint.

$$T = C \cdot U \cdot S$$

9.5 Beyond the Normalized Constraint

Given the set of invariants averages invariance and normalization.

$$\begin{cases} p_1 + p_2 + \cdots + p_n = 1 \\ p_1 F_1 + p_2 F_2 + \cdots + p_n F_n = \langle F \rangle \\ \quad \cdots \\ p_1 f_1 + p_2 f_2 + \cdots + p_n f_n = \langle f \rangle \end{cases} \tag{9.58}$$

We have the Jacobian

$$J = \begin{bmatrix} 1 & F_1 & \cdots & f_1 \\ 1 & F_2 & \cdots & f_2 \\ \cdots & \cdots & \cdots & \cdots \\ 1 & F_n & \cdots & f_n \end{bmatrix}$$

And the genetic evolution subject to constraint is (9.59)

$$p(t+1) = Tp(t)(J^T Tp(t))^{-1} J^T \tag{9.59}$$

9.6   Conclusion

We applied morphogenetic to genetic algorithm which provides a new method to address the core issues in GA. Using projection theorem, morphogenetic first gives the direction of the search which is able to address the convergence problem, and then, it starts the search for optimal solutions from the global perspective which would avoid to mistake local optimal solutions for global optimal solutions.

Chapter 10
Neural Morphogenetic Computing and One Step Method

10.1 One Step Back Propagation to Design Neural Network

Systems design is the process of defining the architecture, components, modules, interfaces, and data for a system to satisfy specified requirements [11, 14, 15]. One step back propagation is a method to design a neural network to satisfy specific Boolean function in output. This is and was the main problem in a percetron neural system and in the classical back propagation. In the classical learning process in neural network we begin with a random values of the parameters and with an iteration program we compute the neural parameters to learn the wanted Boolean function. With the learning program the neural network evolve in time to the wanted goal. Different algorithm have different degrees of evolvability or number of steps to calculate the parameters to satisfy specific Boolean function. Specific programs and functions have different degrees of evolvability. We know that with a given input in the neuron we have Boolean functions where the number of steps have no limit and the evolvability is zero. Other function can be solved by parameters in only one step so the evolvability is one. When the evolvability is zero back propagation in an empirical way adjoin new hidden variables and layers to solve the Boolean function. Because in back propagation we use the descendent gradient to control the evolution of the system when the gradient is zero we have a singularity in the method for which we are trapped in a local minimum. In this case the system converges to a Boolean that does not satisfy the specific requirements. In this chapter we suggest a one step method without the evolution process. Given a set of N inputs to the neuron whose values are one and zero. We have different evaluations for the inputs. The first type of evaluation gives the basis sets of inputs or factors where only one input assumes the value one and all the other values zero. So this basis type of inputs are N and the independent set of inputs or factors. Given the basis set of N inputs with a superposition operation of different independent set

© Springer International Publishing AG 2017
G. Resconi et al., *Introduction to Morphogenetic Computing*,
Studies in Computational Intelligence 703, DOI 10.1007/978-3-319-57615-2_10

of inputs we can generate many other inputs that are dependent from the basis elements as in the coordinate system of the basis vectors of N dimensions.

We compute the weights of the connections and the hidden units which are not part of the input or output in one step. We know neural networks classical back propagation can learn their weights and biases using the gradient descent algorithm unsupervised and the cost function. In this chapter the one step back propagation uses pure algebraic methods to compute the neuron parameters where the cost function and gradient are not useful. We compute again the neuron parameters as in the classical back-propagation and simpler methods such as the perceptron-convergence procedure but without convergent procedure with only algebraic one step algorithm. With the new one step method we can also compute the parameters for associative memory, Hopfield neural network, Kohonen self organizing maps and pattern memory. The one step back propagation defines a Boolean vector of 1 and 0 values as output and a set of Boolean vectors as input. The input set of the vectors are part of a multidimensional vector space and are the reference that we use in the one step method. The reference given by the input vectors is in general non orthogonal and the vectors of the reference are not unitary vectors. In this method we use massive parallel process where all possible inputs which values are one or zero are given by a set of vectors in a multidimensional space. Tor two inputs the input space has dimension 4 for thee input the dimensions space has 8 dimensions and so on. We remember that the number of Boolean function that we want to implement by neural network for two inputs are 16, for three inputs are 256 for four inputs are 65,536. The number of the Boolean function grows up more than the dimension of the space. Now the designed Boolean function that we want to implement by the neural network is the initial part of the algorithm and back we want to use the input space to compute the weights and threshold of the neuron. The neuron parameters are computed without a recursive method of approximation but with a simple use of the vector algebra. At the first the algorithm use a special operator denoted projection operator that can project the output vector into the vector space of the possible inputs. Now the values of the general coordinates in the special input space are the weights of the neuron. Given the weights we can compute the output vector by the weighted linear combination of the input vectors. The output is a general vector which value are numbers. To rebuilt the designed output function we compute the threshold value by the average of the maximum value for output which designed value is zero and the minimum value of all the values which designed value is one. In many case the parameters and the threshold cannot compute the designed Boolean function. When we cannot rebuilt the designed function we introduce new hidden layers in this way.

First this chapter begins with back propagation whose weights and bias are computed with a new method denoted projection methods or one step method. Then we compute with the same method the associative memory parameters. Associative memory is a good example of the rules and data fusion. In the classical associative memory we use orthonormal property of the input samples. Now we show that this property is not always necessary. So we move from associative memory to propagators. After the associative memory we study self associative memory as Hopfield neural network. The unsupervised Kohonen self-organizing maps is included in projection method as special cases. Aims or intentions can be represented by special samples, Boolean functions, images, input vectors or other systems that we want to implement in the physical model of the brain which is a special part of the universe where rules can change.

10.2   Supervised Neural Network by Projection Method [17–19, 21]

Given one neuron with two inputs and one output, shown in Fig. 10.1.

We compute the weights and threshold of the neuron. Given A the input vectors, w the weights, and Y the desired vector,

$$A = \begin{bmatrix} 0 & 0 \\ 1 & 0 \\ 0 & 1 \\ 1 & 1 \end{bmatrix}, Y = \begin{bmatrix} 0 \\ 1 \\ 0 \\ 0 \end{bmatrix},$$

we have Eq. (10.1).

$$Aw = Y \tag{10.1}$$

or

$$\begin{bmatrix} 0 & 0 \\ 1 & 0 \\ 0 & 1 \\ 1 & 1 \end{bmatrix} \begin{bmatrix} w_1 \\ w_2 \end{bmatrix} = \begin{bmatrix} 0 \\ 1 \\ 0 \\ 0 \end{bmatrix}$$

What we want to solve is to get the weights by the input and desired function Y. According to the previous equation, we get the weights w by (10.2).

Fig. 10.1 The neuron with two inputs and one output

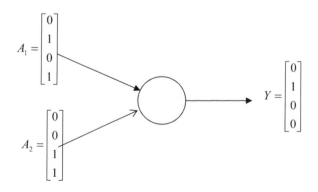

$$A^T A w = A^T Y,$$
$$w = (A^T A)^{-1} A^T Y \tag{10.2}$$

And

$$w = (A^T A)^{-1} A^T Y = \begin{bmatrix} \frac{2}{3} \\ -\frac{1}{3} \end{bmatrix}$$

With w we can compute the projection QY of Y into the input world A by (10.3).

$$A w = A(A^T A)^{-1} A^T Y = QY \tag{10.3}$$

$$QY = A w = \begin{bmatrix} 0 & 0 \\ 1 & 0 \\ 0 & 1 \\ 1 & 1 \end{bmatrix} \begin{bmatrix} \frac{2}{3} \\ -\frac{1}{3} \end{bmatrix} = \frac{2}{3} \begin{bmatrix} 0 \\ 1 \\ 0 \\ 1 \end{bmatrix} - \frac{1}{3} \begin{bmatrix} 0 \\ 1 \\ 0 \\ 1 \end{bmatrix} = w_1 A_1 + w_2 A_2 = \begin{bmatrix} 0 \\ \frac{2}{3} \\ -\frac{1}{3} \\ \frac{1}{3} \end{bmatrix}$$

$$\tag{10.4}$$

For the projection operator the linear combination of the column vectors in (10.4) assumes the minimal value of difference QY-Y among all possible linear combinations.

Because the neuron-like neuron can be given biases by introducing an extra input to each unit which always has a value of 1, the weight θ on this extra input is called the bias and is equivalent to a threshold of the opposite sign. To compute the threshold we use the expression (10.5).

$$\theta = \frac{\min[(QY)Y] + \max[(QY)(1-Y)]}{2} \tag{10.5}$$

In the previous example we have

$$\theta = \frac{\min[0, \frac{2}{3}, 0, 0] - \max[0, 0, -\frac{1}{3}, \frac{1}{3}]}{2} + \max[0, 0, -\frac{1}{3}, \frac{1}{3}] = \frac{\frac{2}{3} - \frac{1}{3}}{2} + \frac{1}{3} = \frac{1}{2}$$

Calling the output y_j we can write (10.6).

$$y_j = f[\sum_i w_i X_{i,j} - \theta] \tag{10.6}$$

Where f is the step function (actually known as the Heviside function) and

$$f(x) = 1 \quad x > 0$$
$$f(x) = 0 \quad x \le 0$$

In this situation no hidden neurons are necessary. With projection operator we find the strength of the connections with the desired output without iteration process. Among the eight desired functions Y in (10.7), only

$$Y = \begin{bmatrix} 0 \\ 1 \\ 1 \\ 0 \end{bmatrix}$$

cannot be solved by projection operator.

$$Y = \begin{bmatrix} 0 \\ 0 \\ 0 \\ 0 \end{bmatrix}, \begin{bmatrix} 0 \\ 1 \\ 0 \\ 0 \end{bmatrix}, \begin{bmatrix} 0 \\ 0 \\ 1 \\ 0 \end{bmatrix}, \begin{bmatrix} 0 \\ 0 \\ 0 \\ 1 \end{bmatrix}, \begin{bmatrix} 0 \\ 1 \\ 1 \\ 0 \end{bmatrix}, \begin{bmatrix} 0 \\ 1 \\ 0 \\ 1 \end{bmatrix}, \begin{bmatrix} 0 \\ 0 \\ 1 \\ 1 \end{bmatrix}, \begin{bmatrix} 0 \\ 1 \\ 1 \\ 1 \end{bmatrix} \tag{10.7}$$

Now,

$$A = \begin{bmatrix} 0 & 0 \\ 1 & 0 \\ 0 & 1 \\ 1 & 1 \end{bmatrix}, Y = \begin{bmatrix} 0 \\ 1 \\ 1 \\ 0 \end{bmatrix}$$

This is a XOR Boolean function. We compute QY by the projection operator (10.8).

$$QY = (A(A^TA)^{-1}A^T)Y = \begin{bmatrix} 0 \\ \frac{1}{3} \\ \frac{1}{3} \\ \frac{2}{3} \end{bmatrix} \quad \text{and} \quad Q = A(A^TA)^{-1}A^T \tag{10.8}$$

So in this case it is impossible to solve the neuron problem.

10.3 Conflict Situation in Supervised Neural Network with Compensation

In the previous problem we have for A and Y we have for the XOR the matrices.

$$A = \begin{bmatrix} 0 & 0 \\ 1 & 0 \\ 0 & 1 \\ 1 & 1 \end{bmatrix}, Y = \begin{bmatrix} 0 \\ 1 \\ 1 \\ 0 \end{bmatrix}$$

Where we have only two free variables, and the other are zero or linear combinations of the two free variables. In fact we have (10.9).

$$Aw = \begin{bmatrix} 0 & 0 \\ 1 & 0 \\ 0 & 1 \\ 1 & 1 \end{bmatrix} \begin{bmatrix} w_1 \\ w_2 \end{bmatrix} = \begin{bmatrix} 0 \\ w_1 \\ w_2 \\ w_1 + w_2 \end{bmatrix} \tag{10.9}$$

So we can put

$$w = \begin{bmatrix} w_1 \\ w_2 \end{bmatrix} = \begin{bmatrix} 1 \\ 1 \end{bmatrix}$$

And

$$Aw = \begin{bmatrix} 0 & 0 \\ 1 & 0 \\ 0 & 1 \\ 1 & 1 \end{bmatrix} \begin{bmatrix} 1 \\ 1 \end{bmatrix} = \begin{bmatrix} 0 \\ 1 \\ 1 \\ 2 \end{bmatrix}$$

When we compare with the original Y we have (10.10)

$$Aw - Y = \begin{bmatrix} 0 & 0 \\ 1 & 0 \\ 0 & 1 \\ 1 & 1 \end{bmatrix} \begin{bmatrix} 1 \\ 1 \end{bmatrix} - \begin{bmatrix} 0 \\ 1 \\ 1 \\ 0 \end{bmatrix} = \begin{bmatrix} 0 \\ 0 \\ 0 \\ 2 \end{bmatrix} \tag{10.10}$$

The last value of the difference is 2 and is the contradiction term for which we cannot solve the neuron problem. In a more simple way we show the contradiction directly. Given the abstract form of the neuron composition in this form

$$\begin{bmatrix} 0 & 0 \\ 1 & 0 \\ 0 & 1 \\ 1 & 1 \end{bmatrix} \begin{bmatrix} w_1 \\ w_2 \end{bmatrix} = \begin{bmatrix} 0 \\ w_1 \\ w_2 \\ w_1 + w_2 \end{bmatrix}$$

For which we have the possible output

$$\begin{bmatrix} 0 \\ w_1 - \theta \\ w_2 - \theta \\ w_1 + w_2 - \theta \end{bmatrix} = \begin{bmatrix} 0 \\ Y_1 \\ Y_2 \\ Y_1 + Y_2 + \theta \end{bmatrix}$$

But for the XOR function we have the output $Y = \begin{bmatrix} 0 \\ 1 \\ 1 \\ 0 \end{bmatrix}$ But this generates the

conflict because

$$(Y_1 > 0, Y_2 > 0) \rightarrow (Y_1 + Y_2) > 0$$

And this is in contradiction with the Boolean constraint in XOR function for which

$$(Y_1 + Y_2) + \theta < 0$$

For $Y_1 + Y_2 > 0$ and $\theta > 0$ the previous expression is impossible.
So we adjoin a new column with 1 in this point where we have contradiction. In conclusion with the new input we have the network (10.11).

$$A' = \begin{bmatrix} 0 & 0 & 0 \\ 1 & 0 & 0 \\ 0 & 1 & 0 \\ 1 & 1 & 1 \end{bmatrix}, Y = \begin{bmatrix} 0 \\ 1 \\ 1 \\ 0 \end{bmatrix} \tag{10.11}$$

Where we have three free variables with $w = \begin{bmatrix} 1 \\ 1 \\ -2 \end{bmatrix}, \theta = 0.5, QY = \begin{bmatrix} 0 \\ 1 \\ 1 \\ 0 \end{bmatrix}$ for

$Y = \begin{bmatrix} 0 \\ 0 \\ 0 \\ 1 \end{bmatrix}$ and $A = \begin{bmatrix} 0 & 0 \\ 1 & 0 \\ 0 & 1 \\ 1 & 1 \end{bmatrix}$ with the projection method we have $w = \begin{bmatrix} \frac{1}{3} \\ \frac{1}{3} \\ \frac{1}{3} \end{bmatrix}$ and the

threshold 0.5, So $QY = \begin{bmatrix} 0 \\ 1/3 \\ 1/3 \\ 2/3 \end{bmatrix}$

We can build this neural network (Fig. 10.2).

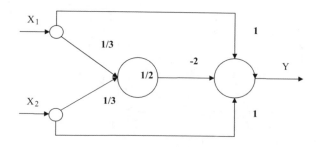

Fig. 10.2 Weights and bias for the Boolean function XOR by the projection method without cost function and descendent gradient

Example For the three inputs neuron we have (10.12).

$$
A = \begin{bmatrix} 0 & 0 & 0 \\ 1 & 0 & 0 \\ 0 & 1 & 0 \\ 1 & 1 & 0 \\ 0 & 0 & 1 \\ 1 & 0 & 1 \\ 0 & 1 & 1 \\ 1 & 1 & 1 \end{bmatrix}, \ Y = \begin{bmatrix} 0 \\ 1 \\ 1 \\ 0 \\ 1 \\ 0 \\ 0 \\ 1 \end{bmatrix}
\qquad (10.12)
$$

With three free variables, we assume w to be:

$$
w = \begin{bmatrix} w_1 \\ w_2 \\ w_3 \end{bmatrix} = \begin{bmatrix} 1 \\ 1 \\ 1 \end{bmatrix}
$$

And

$$
D = Aw - Y = \begin{bmatrix} 0 \\ 0 \\ 0 \\ 2 \\ 0 \\ 2 \\ 2 \\ 2 \end{bmatrix}
$$

Where D is the difference between Aw and Y. When we adjoin the difference to the input values A, we have that for (10.13).

$$
A = \begin{bmatrix} 0 & 0 & 0 & 0 \\ 1 & 0 & 0 & 0 \\ 0 & 1 & 0 & 0 \\ 1 & 1 & 0 & 2 \\ 0 & 0 & 1 & 0 \\ 1 & 0 & 1 & 2 \\ 0 & 1 & 1 & 2 \\ 1 & 1 & 1 & 2 \end{bmatrix}, Y = \begin{bmatrix} 0 \\ 1 \\ 1 \\ 0 \\ 1 \\ 0 \\ 0 \\ 1 \end{bmatrix}
\qquad (10.13)
$$

We have $w = (A^T A)^{-1} A^T Y = \begin{bmatrix} 1 \\ 1 \\ 1 \\ -1 \end{bmatrix}$ and (10.14)

$$QY = A(A^TA)^{-1}A^TY = \begin{bmatrix} 0 \\ 1 \\ 1 \\ 0 \\ 1 \\ 0 \\ 0 \\ 1 \end{bmatrix} \tag{10.14}$$

The previous basis A can be split in this way.

$$A = \begin{bmatrix} 0 & 0 & 0 & 0 \\ 1 & 0 & 0 & 0 \\ 0 & 1 & 0 & 0 \\ 1 & 1 & 0 & 2 \\ 0 & 0 & 1 & 0 \\ 1 & 0 & 1 & 2 \\ 0 & 1 & 1 & 2 \\ 1 & 1 & 1 & 2 \end{bmatrix} = \begin{bmatrix} 0 & 0 & 0 & 0 & 0 \\ 1 & 0 & 0 & 0 & 0 \\ 0 & 1 & 0 & 0 & 0 \\ 1 & 1 & 0 & 1 & 1 \\ 0 & 0 & 1 & 0 & 0 \\ 1 & 0 & 1 & 1 & 1 \\ 0 & 1 & 1 & 1 & 1 \\ 1 & 1 & 1 & 1 & 1 \end{bmatrix}$$

But because the last two columns are equal we can write the new basis in this way.

$$A = \begin{bmatrix} 0 & 0 & 0 & 0 \\ 1 & 0 & 0 & 0 \\ 0 & 1 & 0 & 0 \\ 1 & 1 & 0 & 1 \\ 0 & 0 & 1 & 0 \\ 1 & 0 & 1 & 1 \\ 0 & 1 & 1 & 1 \\ 1 & 1 & 1 & 1 \end{bmatrix}$$

In (10.15) we have the elements to compute the weights of the neural network. So

$$A = \begin{bmatrix} 0 & 0 & 0 & 0 \\ 1 & 0 & 0 & 0 \\ 0 & 1 & 0 & 0 \\ 1 & 1 & 0 & 1 \\ 0 & 0 & 1 & 0 \\ 1 & 0 & 1 & 1 \\ 0 & 1 & 1 & 1 \\ 1 & 1 & 1 & 1 \end{bmatrix}, \quad Y = \begin{bmatrix} 0 \\ 1 \\ 1 \\ 0 \\ 1 \\ 0 \\ 0 \\ 1 \end{bmatrix} \tag{10.15}$$

The weights and the projection do not change. So we have again the same result.

We have $w = (A^T A)^{-1} A^T Y = \begin{bmatrix} 1 \\ 1 \\ 1 \\ -1 \end{bmatrix}$ and $QY = A(A^T A)^{-1} A^T Y = \begin{bmatrix} 0 \\ 1 \\ 1 \\ 0 \\ 1 \\ 0 \\ 0 \\ 1 \end{bmatrix}$

Now we want to solve by neural network the function.

$$Y = \begin{bmatrix} 0 \\ 0 \\ 0 \\ 1 \\ 0 \\ 1 \\ 1 \\ 1 \end{bmatrix} \qquad (10.16)$$

With the projection operator we have

$$w = \frac{3}{8} \begin{bmatrix} 1 \\ 1 \\ 1 \end{bmatrix} \text{ and } \theta = 0.563$$

So we have the result as Fig. 10.3.
We show the neural network parameters in Fig. 10.4.
Now we show another example. For the three input neuron we have

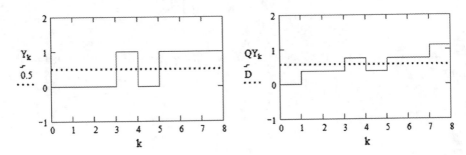

Fig. 10.3 Function Y and function of the projection or QY. We see that the two functions with the bias are equal

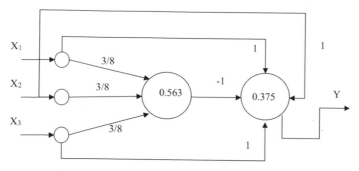

Fig. 10.4 Neural network with weights and bias for the function 01101001 with projection method

$$
A = \begin{bmatrix} 0 & 0 & 0 \\ 1 & 0 & 0 \\ 0 & 1 & 0 \\ 1 & 1 & 0 \\ 0 & 0 & 1 \\ 1 & 0 & 1 \\ 0 & 1 & 1 \\ 1 & 1 & 1 \end{bmatrix}, \; Y = \begin{bmatrix} 0 \\ 0 \\ 1 \\ 1 \\ 0 \\ 1 \\ 0 \\ 1 \end{bmatrix}
\tag{10.17}
$$

With projection method we have Fig. 10.5.

So QY cannot solve the neuron problem with $w = \begin{bmatrix} 0.5 \\ 0.5 \\ 0 \end{bmatrix}, \theta = 0.5$

So we must adjoin a new column. For $w = \begin{bmatrix} w_1 \\ w_2 \\ w_3 \end{bmatrix} = \begin{bmatrix} 1 \\ 1 \\ 1 \end{bmatrix}$ we have

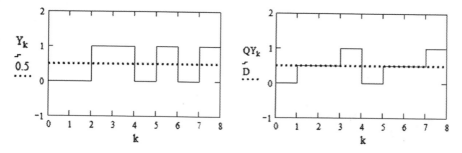

Fig. 10.5 Projection method for the Boolean function 00111010 the QY cannot obtain the wanted function Y

$$D = Aw - Y = \begin{bmatrix} 0 \\ 1 \\ 0 \\ 1 \\ 1 \\ 1 \\ 2 \\ 2 \end{bmatrix}$$

Where D is the difference between Aw and Y. When we adjoin the difference to the input values A, we have that for

$$A = \begin{bmatrix} 0 & 0 & 0 & 0 \\ 1 & 0 & 0 & 1 \\ 0 & 1 & 0 & 0 \\ 1 & 1 & 0 & 1 \\ 0 & 0 & 1 & 1 \\ 1 & 0 & 1 & 1 \\ 0 & 1 & 1 & 2 \\ 1 & 1 & 1 & 2 \end{bmatrix}, \quad Y = \begin{bmatrix} 0 \\ 0 \\ 1 \\ 1 \\ 0 \\ 1 \\ 0 \\ 1 \end{bmatrix}$$

We have $w = (A^T A)^{-1} A^T Y = \begin{bmatrix} 1 \\ 1 \\ 1 \\ -1 \end{bmatrix}$ and $QY = A(A^T A)^{-1} A^T Y = \begin{bmatrix} 0 \\ 0 \\ 1 \\ 1 \\ 0 \\ 1 \\ 0 \\ 1 \end{bmatrix}$

So we solve the neuron problem. Now because the projection operator is a linear weighted composition of the columns of A, when we split A in two parts the linear combination cannot change and we can solve the neuron in the same way. So for D we have the two columns.

$$D = \begin{bmatrix} 0 \\ 1 \\ 0 \\ 1 \\ 1 \\ 1 \\ 2 \\ 2 \end{bmatrix} \rightarrow \begin{bmatrix} 0 & 0 \\ 1 & 0 \\ 0 & 0 \\ 1 & 0 \\ 1 & 0 \\ 1 & 0 \\ 1 & 1 \\ 1 & 1 \end{bmatrix}$$

For the property of the projection operator we have the new basis A.

$$A = \begin{bmatrix} 0 & 0 & 0 & 0 & 0 \\ 1 & 0 & 0 & 1 & 0 \\ 0 & 1 & 0 & 0 & 0 \\ 1 & 1 & 0 & 1 & 0 \\ 0 & 0 & 1 & 1 & 0 \\ 1 & 0 & 1 & 1 & 0 \\ 0 & 1 & 1 & 1 & 1 \\ 1 & 1 & 1 & 1 & 1 \end{bmatrix}, \quad Y = \begin{bmatrix} 0 \\ 0 \\ 1 \\ 1 \\ 0 \\ 1 \\ 0 \\ 1 \end{bmatrix}$$

The weights and the projection do not change. So we have again the same result.

We have $w = (A^T A)^{-1} A^T Y = \begin{bmatrix} 1 \\ 1 \\ 1 \\ -1 \\ -1 \end{bmatrix}, \theta = 0.5$ and

$$QY = A(A^T A)^{-1} A^T Y = \begin{bmatrix} 0 \\ 0 \\ 1 \\ 1 \\ 0 \\ 1 \\ 0 \\ 1 \end{bmatrix}$$

Now for the two new columns we have

$$A = \begin{bmatrix} 0 & 0 & 0 \\ 1 & 0 & 0 \\ 0 & 1 & 0 \\ 1 & 1 & 0 \\ 0 & 0 & 1 \\ 1 & 0 & 1 \\ 0 & 1 & 1 \\ 1 & 1 & 1 \end{bmatrix}, \quad Y = \begin{bmatrix} 0 \\ 1 \\ 0 \\ 1 \\ 1 \\ 1 \\ 1 \\ 1 \end{bmatrix}$$

With projection method we have Fig. 10.6.

So QY cannot solve the neuron problem with $w = \begin{bmatrix} 0.5 \\ 0.5 \\ 0.5 \end{bmatrix}, \theta = 0.25$.

For the second ne column we have

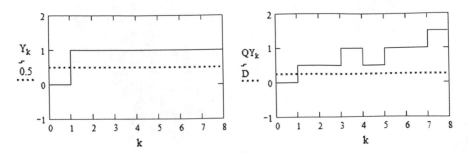

Fig. 10.6 Projection method for the first new column

$$A = \begin{bmatrix} 0 & 0 & 0 \\ 1 & 0 & 0 \\ 0 & 1 & 0 \\ 1 & 1 & 0 \\ 0 & 0 & 1 \\ 1 & 0 & 1 \\ 0 & 1 & 1 \\ 1 & 1 & 1 \end{bmatrix}, Y = \begin{bmatrix} 0 \\ 0 \\ 0 \\ 0 \\ 0 \\ 0 \\ 1 \\ 1 \end{bmatrix}$$

With projection method we have Fig. 10.7.

So QY cannot solve the neuron problem with $w = \frac{1}{8} \begin{bmatrix} -1 \\ 3 \\ 3 \end{bmatrix}, \theta = 0.5$.

So we have the network for the Boolean function 00110101 (Fig. 10.8).

So we have the maximum of the stability QY = Y. Now when we reduce the number of columns we have

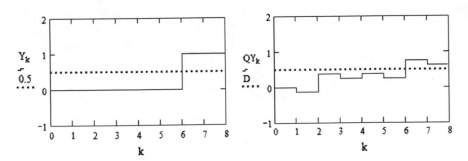

Fig. 10.7 Projection method for the second new column

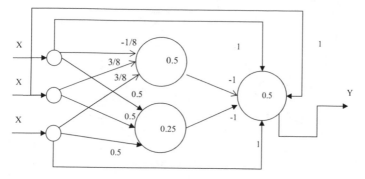

Fig. 10.8 Neural network for the Boolean function 00110101

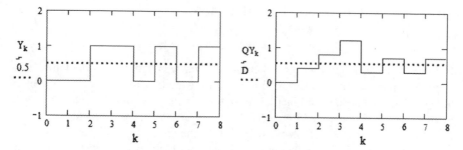

Fig. 10.9 Solution of the function 00110101 with the projection method with only one new column. QY is different from Y but with the bias we can have that QY = Y and we solve the neural network problem

$$
A = \begin{bmatrix}
0 & 0 & 0 & 0 \\
1 & 0 & 0 & 0 \\
0 & 1 & 0 & 0 \\
1 & 1 & 0 & 0 \\
0 & 0 & 1 & 0 \\
1 & 0 & 1 & 0 \\
0 & 1 & 1 & 1 \\
1 & 1 & 1 & 1
\end{bmatrix}
$$

With the projection method we have Fig. 10.9.

With $w = \frac{1}{5} \begin{bmatrix} 2 \\ 4 \\ 6 \\ -4 \end{bmatrix}, \theta = 0.5$.

So we have only one neuron but with less stability and less neurons. The neural network is (Fig. 10.10).

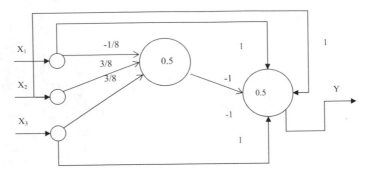

Fig. 10.10   Neural network with only new hidden neuron to solve the Boolean function 00110101

10.4   Evolvability and One Step Method in Neural Network

In the classical learning process in neural network we begin with a random values of the parameters and with an iteration program we compute the neural parameters to learn the wanted Boolean function. With the learning program the neural network evolve in time to the wanted goal. Different algorithms have different degrees of evolvability or number of steps to calculate the parameters to satisfy specific Boolean function. We know that with a given input in the neuron we have Boolean functions where the number of steps have no limit and the evolvability is zero. Other functions can be solved by parameters in only one step so the evolvability is one. When the evolvability is zero back propagation in an empirical way adjoins new hidden variables and layers to solve the Boolean function. Because in back-propagation we use the descendent gradient to control the evolution of the system when the gradient is zero we have a singularity in the method for which we are trapped in a local minimum. In this case the system converges to a Boolean that does not satisfy to the specific requirements. In this chapter we suggest a one step method without the evolution process. The neuron has N inputs whose values are one and zero. The sets of the inputs whose value is one are the subsets of the all possible inputs to the neurons. The empty set is the input where all the inputs are equal to zero. The subsets of the inputs with only one element equal to one and all the others equal to zero are the basic or elements of the input values. All the other subsets are the union of the basic subsets. Because the number of the subsets for a set with N elements are 2^N this is the number of all possible types of inputs where any input is a vector of N elements whose value is one or zero. Now we can transform any Boolean input vector in one integer number and for the number order 0, 1, 2, ..., $2^N - 1$. we can create a matrix A of ordered inputs. At any individual input we associate a column vector whose dimension is 2^N and at any set of inputs at the same time we associate a row vector with dimension N. The matrix M is a set non orthogonal column vectors that we denote as the column space. Now given the vector space of the columns the linear combinations of the column vectors generate all the space. Given a specific Boolean vectors V we search the best weights of the linear combination of the column vectors and by threshold process give back the specific function V. If we find the weights we find the relation between the inputs to the specific Boolean function in the neuron. The algorithm of the projection makes it possible give the result. We can prove that if the projection method can find by vector calculus the solution of the neuron problem, any previous methods that use recursion methods can find the same result. The difference is that the projection method is one step method and all the others are many steps methods. When the projection method cannot solve the problem that the evolvability assumes the value zero, we cannot solve the problem in any case and with any method (percetron problem). As in the classical back propagation we must adjoin hidden layer to solve the problem, to solve this problem we transform the set of column vector into integer numbers P obtained by the superposition of the column vectors with the weights all equal to one. After we subtract from the vector P the vector V, the new vector D is adjoined to the previous input to obtain N + 1 column vectors. The new column space includes the vector V as a linear combination of the N + 1 vectors. Now we split the vector D in a set of vectors with value one and zero. So now the column space is N + h that solves again the neuron problem. When we delete part of the new columns we eliminate layers for which we can again solve the neuron problem with threshold value but with less and less of stability. In conclusion now with the projection method we compute the layers in back propagation in a way to transform a non-evolvable system to evolvable with different degree of stability. At the same time with algebraic one step method we can compute the wanted neural parameters. We begin with a very simple neural system to show the one step method.

$$Aw = Y$$
$$A^T Aw = A^T Y,$$
$$w = (A^T A)^{-1} A^T Y$$
$$Aw = A(A^T A)^{-1} A^T Y = QY$$

We can show that QY is the projection of Y into the column space of the rectangular matrix A. Because the neuron-like neuron can be given biases by introducing an extra input to each unit which always has a value of 1, the weight θ on this extra input is called the bias and is equivalent to a threshold of the opposite sign. To compute the threshold we use the expression (10.18).

$$\theta = \frac{\min[(QY)Y] + \max[(QY)(1 - Y)]}{2}$$
$$y_j = f\left[\sum_i w_i X_{i,j} - \theta\right]$$

(10.18)

Given the matrix A and the Y as the specific function we take for the weights w as all equal to one so we have the vector D

Table 10.1 Neuron input and output vectors

	Neuron input 1	Neuron input 2	Neuron input 3	Neuron input 4	Neuron output
State 0	0	0	0	0	0
State 1	1	0	0	0	1
State 2	0	1	0	0	0
State 3	1	1	0	0	1
State 4	0	0	1	0	0
State 5	1	0	1	0	1
State 6	0	1	1	0	0
State 7	1	1	1	0	1
State 8	0	0	0	1	0
State 9	1	0	0	1	1
State 10	0	1	0	1	0
State 11	1	1	0	1	1
State 12	0	0	1	1	1
State 13	1	0	1	1	0
State 14	0	1	1	1	1
State 15	1	1	1	1	1

$$D = Aw - Y$$

That is the new internal input or layer to adjoin at the matrix A of all the possible inputs. Table 10.1 shows input and output vectors.

When A is the matrix of the input with four columns and 16 rows and Y is the assigned output Y, we compute the weights $w = [0.65, 0.15, 0, 15, 0.15]$ and the threshold value is $\theta = 0.625$. The output values can be compared with the computed values so we have the results (Fig. 10.11).

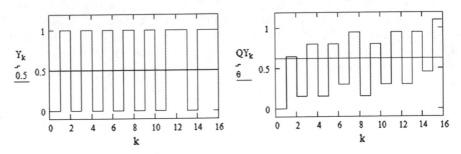

Fig. 10.11 Function Y and projection QY. The projection operator cannot generate the same function Y

Table 10.2 Neuron input and output and layer D vectors

	Neuron input 1	Neuron input 2	Neuron input 3	Neuron input 4	Neuron input D	Neuron output
State 0	0	0	0	0	0	0
State 1	1	0	0	0	0	1
State 2	0	1	0	0	1	0
State 3	1	1	0	0	1	1
State 4	0	0	1	0	1	0
State 5	1	0	1	0	1	1
State 6	0	1	1	0	2	0
State 7	1	1	1	0	2	1
State 8	0	0	0	1	1	0
State 9	1	0	0	1	1	1
State 10	0	1	0	1	2	0
State 11	1	1	0	1	2	1
State 12	0	0	1	1	1	1
State 13	1	0	1	1	3	0
State 14	0	1	1	1	2	1
State 15	1	1	1	1	3	1

We can see that with only Eqs. (10.1), (10.2) and (10.3) we can find the assigned function in the Table 10.1. So we adjoin new inputs to the neuron to solve the problem. We adjoin a column to the previous table so we have Table 10.2.

For the previous inputs we have $Aw = Y$ where A is the matrix of the five inputs and $w = \begin{bmatrix} 1 & 1 & 1 & 1 & -1 \end{bmatrix}$. Now we can split the input D in three inputs with the one zero Boolean values. So we have Table 10.3.

The input 5 can be obtained by the inputs 1, 2, 3, 4 with the weights $w = \begin{bmatrix} 0.2 & 0.45 & 0.45 & 0.45 \end{bmatrix}, \theta = 0.325$ the graph is (Fig. 10.12).

The input 6 cannot solve with the inputs 1, 2, 3, 4 but with the expression (10.6) we see that D = input 7 so we have the inputs 1, 2, 3, 4, 7 for which we have the weights $w = \begin{bmatrix} -0.063 & 0.542 & 0.188 & 0.188 & 0.417 \end{bmatrix}, \theta = 0.604$ and the result is (Fig. 10.13).

For the output 7 can solve with the inputs 1, 2, 3, 4 with the weights $w = \begin{bmatrix} 0.15 & -0.1 & 0.15 & 0.15 \end{bmatrix}, \theta = 0.325$ the result is (Fig. 10.14).

For the previous inputs we have $Aw = Y$ where A is the matrix of the five inputs and $w = \begin{bmatrix} 1 & 1 & 1 & 1 & -1 & -1 & -1 \end{bmatrix}, \theta = 0.5$ (Fig. 10.15).

Table 10.3 Neuron input and output and D expansion vectors

	Neuron input 1	Neuron input 2	Neuron input 3	Neuron input 4	Neuron input 5	Neuron input 6	Neuron input 7	Neuron output
State 0	0	0	0	0	0	0	0	0
State 1	1	0	0	0	0	0	0	1
State 2	0	1	0	0	1	0	0	0
State 3	1	1	0	0	1	0	0	1
State 4	0	0	1	0	1	0	0	0
State 5	1	0	1	0	1	0	0	1
State 6	0	1	1	0	1	1	0	0
State 7	1	1	1	0	1	1	0	1
State 8	0	0	0	1	1	0	0	0
State 9	1	0	0	1	1	0	0	1
State 10	0	1	0	1	1	1	0	0
State 11	1	1	0	1	1	1	0	1
State 12	0	0	1	1	1	0	0	1
State 13	1	0	1	1	1	1	1	0
State 14	0	1	1	1	1	1	0	1
State 15	1	1	1	1	1	1	1	1

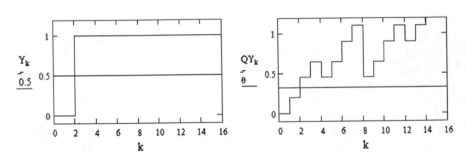

Fig. 10.12 Comparison from the 5 column and the same computed by 1, 2, 3, 4 inputs

When we eliminate the column 5 and 6 we have the inputs 1, 2, 3, 4, 7 so we have the results $w = [0.781 \quad 0.063 \quad 0.281 \quad 0.281 \quad -0.875], \theta = 0.5$ and the result is (Fig. 10.16).

The neurons networks are (Figs. 10.17, 10.18 and 10.19).

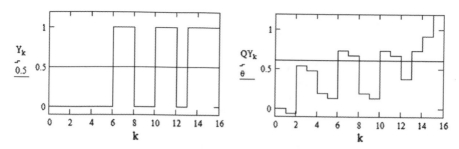

Fig. 10.13 Comparison from the 6 column and the same computed by 1, 2, 3, 4, 7 inputs

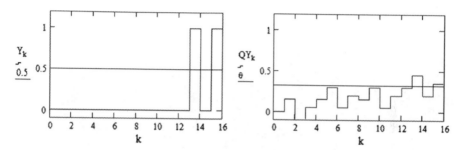

Fig. 10.14 Comparison from the 7 column and the same computed by 1, 2, 3, 4 inputs

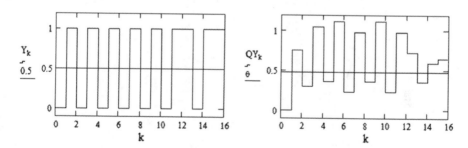

Fig. 10.15 Output by 1, 2, 3, 4, 6, 7 inputs

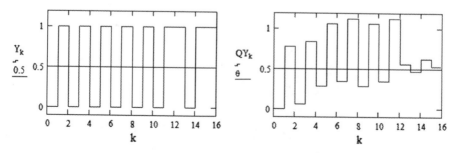

Fig. 10.16 Output by 1, 2, 3, 4, 7 inputs

Fig. 10.17 Neural network
with the inputs 1, 2, 3, 4,
5, 6, 7

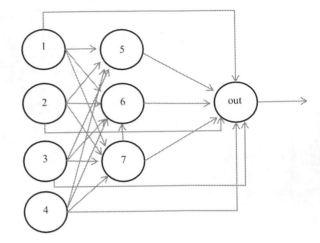

Fig. 10.18 Neural network
with the inputs 1, 2, 3, 4, 6, 7

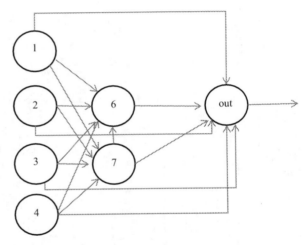

Fig. 10.19 Neural network
with the inputs 1, 2, 3, 4, 7

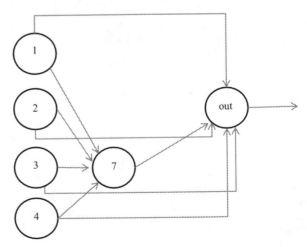

10.5   Associative Memory by One Step Method [5]

At any given point the state of the neural network is given by the activity pattern produced by input and output vectors. Neurons update their activity values based on the inputs they receive (over the synapses). Figure 10.20 shows the neural network with three input vectors and one output vector on the right.

The linear associator is one of the simplest and first studied associative memory model. Figure 10.21 is the network architecture of the linear associator.

In Fig. 10.21, all the m input units are connected to all the n output units via the connection weight matrix $\boldsymbol{W} = [w_{ij}]_{m \times n}$ where w_{ij} denotes the synaptic strength of the unidirectional connection from the ith input unit to the jth output unit.

Given the representation $\overrightarrow{x_k} = [x_{k1}, x_{k2}, \ldots x_{km}]^T$, $\overrightarrow{y_k} = [y_{k1}, y_{k2}, \ldots y_{kn}]^T$, for the jth component y_{kj} $j = 1, 2, \ldots n$, we have (10.19).

$$y_{kj} = [w_{j1}(k), w_{j2}(k), \ldots, w_{jm}(k)] \begin{bmatrix} x_{k1} \\ x_{k2} \\ \ldots \\ x_{km} \end{bmatrix} \tag{10.19}$$

(10.25) can be written in the form of (10.20).

$$y_{kj} = \sum_{i=1}^{m} w_{ji}(k) x_{ki} \tag{10.20}$$

Fig. 10.20 The neural network with neuron-like and synapse-like

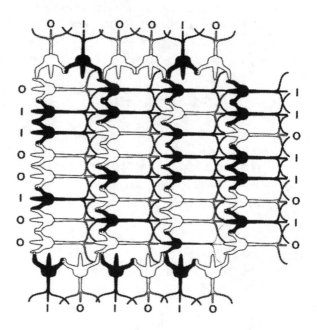

Fig. 10.21 The network of
the linear associator

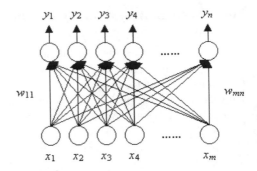

Or (10.21)

$$
\begin{bmatrix} y_{k1} \\ y_{k2} \\ \dots \\ y_{kn} \end{bmatrix} = \begin{bmatrix} w_{11}(k) & w_{12}(k) & \dots & w_{1m}(k) \\ w_{21}(k) & w_{22}(k) & \dots & w_{2m}(k) \\ \dots & \dots & \dots & \dots \\ w_{n1}(k) & w_{n2}(k) & \dots & w_{nm}(k) \end{bmatrix} \begin{bmatrix} x_{k1} \\ x_{k2} \\ \dots \\ x_{km} \end{bmatrix} \tag{10.21}
$$

It is the connection weight matrix W that stores the q different associated pattern pairs $\{(x_k, y_k) \mid k = 1, 2, \dots, q\}$. For every associative pattern k, $\vec{x_k} \rightarrow \vec{y_k}$, We have (10.22).

$$
\vec{y_k} = \overrightarrow{w(k)}\,\vec{x_k} \tag{10.22}
$$

Building an associative memory is nothing but constructing the connection weight matrix W that when an input pattern is presented, the stored pattern associated with the input pattern is retrieved. So we call the set of components W correlation memory matrix $\vec{M}$, which can be represented as (10.23).

$$
\vec{M} = \sum_{k=1}^{q} \overrightarrow{w(k)} \tag{10.23}
$$

So we have $\vec{Y} = \vec{M}\vec{X}$.

The process of constructing the connection weight matrix is called encoding. During encoding the weight values of the correlation matrix $\vec{M}$ for a particular associated pattern pair (x_k, y_k) are computed as (10.24).

$$
W = \sum \vec{y_k}\vec{x_k^T} \tag{10.24}
$$

Or it can be also written like (10.25).

$$\vec{M} = [\vec{y_1}, \vec{y_2}, \dots, \vec{y_q}] \begin{bmatrix} \vec{x_1^T} \\ \vec{x_2^T} \\ \dots \\ \vec{x_q^T} \end{bmatrix} = \vec{Y}\,\vec{X^T} \qquad (10.25)$$

So $\vec{Y} = \vec{M}\,\vec{X} = \vec{Y}\,\vec{X^T}\,\vec{X}$

And $\vec{X^T}\,\vec{X} = $ Identity.

For   example,   if $X = \begin{bmatrix} x_{11} & x_{12} & x_{13} \\ x_{21} & x_{22} & x_{23} \\ \dots & \dots & \dots \\ x_{n1} & x_{n2} & x_{n3} \end{bmatrix}$,   and $\varphi_1 = \begin{bmatrix} x_{11} \\ x_{21} \\ \dots \\ x_{n1} \end{bmatrix}$,

$\varphi_2 = \begin{bmatrix} x_{12} \\ x_{22} \\ \dots \\ x_{n2} \end{bmatrix}, \varphi_3 = \begin{bmatrix} x_{13} \\ x_{23} \\ \dots \\ x_{n3} \end{bmatrix}$,   X   is   represented $X = [\varphi_1 \quad \varphi_2 \quad \varphi_3]$.   Since

$\vec{X^T}\,\vec{X} = $ Identity,

$$\begin{bmatrix} \varphi_1 \\ \varphi_2 \\ \varphi_3 \end{bmatrix} [\varphi_1 \quad \varphi_2 \quad \varphi_3] = Identity$$

So we have $\varphi_i \varphi_i = 1$ and $\varphi_i \varphi_j = 0$ $i \neq j$, which means if the input patterns are mutually orthogonal, perfect retrieval can happen.

The associative memory is a propagator that propagates the information form input X to output Y with the rule

$$Y = WX.$$

Now given the sample $Y_k = WX_k$, if $Y_k = A, X_k = B$, we have $A = WB$. Since A can be written in form (10.26) by one step matrix method

$$A = A(B^T B)^{-1} B^T B \qquad (10.26)$$

We have (10.27)

$$W = A(B^T B)^{-1} B^T \qquad (10.27)$$

When the input samples B are orthogonal we have (10.28).

$$W = AB^T \qquad (10.28)$$

This is the classical associative expression.

Example 10.1

$$A = \begin{bmatrix} y_{11} & y_{12} \\ y_{21} & y_{22} \\ y_{31} & y_{32} \end{bmatrix}, B = \begin{bmatrix} x_{11} & x_{12} \\ x_{21} & x_{22} \\ x_{31} & x_{32} \end{bmatrix}$$

So we have (10.29) one step matrix method

$$W = A(B^T B)^{-1} B^T = \begin{bmatrix} y_{11} & y_{12} \\ y_{21} & y_{22} \\ y_{31} & y_{32} \end{bmatrix} \left(\begin{bmatrix} x_{11} & x_{12} \\ x_{21} & x_{22} \\ x_{31} & x_{32} \end{bmatrix}^T \begin{bmatrix} x_{11} & x_{12} \\ x_{21} & x_{22} \\ x_{31} & x_{32} \end{bmatrix} \right)^{-1} \begin{bmatrix} x_{11} & x_{12} \\ x_{21} & x_{22} \\ x_{31} & x_{32} \end{bmatrix}^T$$

$$= w_{i,j}$$

$$(10.29)$$

The samples in input B define the invariants or rules to be satisfied in the transformation and the samples in output give us the type of transformation that we want to generate. In fact given the inputs $B = \begin{bmatrix} x_{11} & x_{12} \\ x_{21} & x_{22} \\ x_{31} & x_{32} \end{bmatrix} = \begin{bmatrix} 1 & 1 \\ 2 & 1 \\ 3 & 1 \end{bmatrix}$, we have the input invariance (10.30).

$$(x_{32} - x_{22}) - (x_{22} - x_{12}) = x_{32} - 2x_{22} + x_{12} = 0$$
$$(x_{31} - x_{21}) - (x_{21} - x_{11}) = x_{31} - 2x_{21} + x_{11} = 0 \qquad (10.30)$$

That means the components of the input vectors are in the straight line. Since the samples are points in the straight line, the output vectors are points in the straight line.

Now when the output samples are transformations of the inputs as $A = \Omega B$, We have (10.31).

$$W = \Omega B (B^T B)^{-1} B^T = \Omega B (B^T B)^{-1} B^T \qquad (10.31)$$

The parameters W of the associative memory can be written in (10.32).

$$W = \Omega B (B^T B)^{-1} B^T = \Omega Q \qquad (10.32)$$

Where Q is the projection operator of the vector p into the input space sample B. In fact we have

Fig. 10.22 The association of external vector and its projection

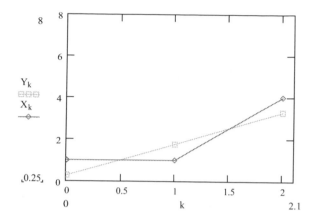

$$Q = B(B^T B)^{-1} B^T$$
$$Q^2 = B(B^T B)^{-1} B^T B(B^T B)^{-1} B^T = Q \qquad (10.33)$$

For A = B we have (10.34).

$$Wy = B(B^T B)^{-1} B^T y = B(B^T B)^{-1} B^T y \qquad (10.34)$$

For $y = \begin{bmatrix} 1 \\ 1 \\ 4 \end{bmatrix}$, We give the association of external vector and its projection as Fig. 10.22. Blue line represents external vector (X_k), and red line is the vector (Y_k) with the straight line property generated by projection operator.

Where the X_k and Y_k are associated. They are not equal because the external vector is not be in agreement with the rules in B that are straight lines.

For A = 2B, we have Fig. 10.23. We also have a red straight line but with the expansion of the coordinate space ($\Omega = 2$).

Example 10.2 If $\Omega = R(\alpha) = \begin{bmatrix} \cos(\alpha) & \sin(\alpha) & 0 \\ -\sin(\alpha) & \cos(\alpha) & 0 \\ 0 & 0 & 1 \end{bmatrix}$ is the rotation operator and

$$B = \begin{bmatrix} x_{11} & x_{12} \\ x_{21} & x_{22} \\ x_{31} & x_{32} \end{bmatrix} = \begin{bmatrix} 1 & 1 \\ 2 & 1 \\ 3 & 1 \end{bmatrix}$$

We have (10.35)

Fig. 10.23 The projection
with the expansion of the
coordinate (colour figure
online)

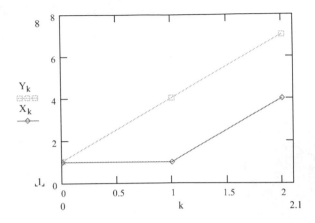

$$W = A(B^T B)^{-1} B^T = \Omega Q$$

$$= \begin{bmatrix} \cos(\alpha) & \sin(\alpha) & 0 \\ -\sin(\alpha) & \cos(\alpha) & 0 \\ 0 & 0 & 1 \end{bmatrix} \begin{bmatrix} x_{11} & x_{12} \\ x_{21} & x_{22} \\ x_{31} & x_{32} \end{bmatrix} \left(\begin{bmatrix} x_{11} & x_{12} \\ x_{21} & x_{22} \\ x_{31} & x_{32} \end{bmatrix}^T \begin{bmatrix} x_{11} & x_{12} \\ x_{21} & x_{22} \\ x_{31} & x_{32} \end{bmatrix} \right)^{-1} \begin{bmatrix} x_{11} & x_{12} \\ x_{21} & x_{22} \\ x_{31} & x_{32} \end{bmatrix}^T$$

$$= \begin{bmatrix} \cos(\alpha) & \sin(\alpha) & 0 \\ -\sin(\alpha) & \cos(\alpha) & 0 \\ 0 & 0 & 1 \end{bmatrix} \begin{bmatrix} q_{11} & q_{12} & q_{13} \\ q_{21} & q_{22} & q_{23} \\ q_{31} & q_{32} & q_{33} \end{bmatrix}$$

$$= \begin{bmatrix} q_{11}\cos(\alpha) + q_{21}\sin(\alpha) & q_{12}\cos(\alpha) + q_{22}\sin(\alpha) & q_{13}\cos(\alpha) + q_{23}\sin(\alpha) \\ -\sin(\alpha)q_{11} + \cos(\alpha)q_{21} & -\sin(\alpha)q_{12} + \cos(\alpha)q_{22} & -\sin(\alpha)q_{13} + \cos(\alpha)q_{23} \\ q_{31} & q_{32} & q_{33} \end{bmatrix}$$

$$(10.35)$$

In the associative memory we compute the weights as an operator (memory) that include invariance. So any association generate in output elements of the same universe with the same rules or invariance. In digital computer we have passive memory in brain or morphogenetic computing we have data but also rules. Any input to the brain is changes in a way to have new data associate to the input with the internal rule at the memory.

10.6   Hopfield Neural Network and Morphogenetic Computing as One Step Method [7]

The Hopfield Network (1982) can be represented by Fig. 10.24.

Where we can see that Hopfield neural network is a self or bidirectional (loop) of neural network. As described above, in the associative memory we have

Fig. 10.24 Hopfield network

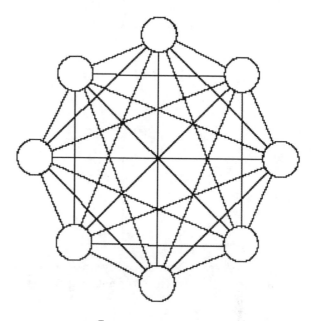

$$W = YX^T$$

In Hopfield neural network X = Y so we have

$$W = XX^T$$

Because

$$X = \Gamma(x_k) = field\ vector$$

Hopfield Network Basis

$$\Gamma_s(x_k) = set\ of\ field\ vectors$$

fields for n neurons (positions x_k) and correlations as projection operators for orthogonal set of fields is (10.36).

$$
\begin{aligned}
w_{i,j} &= \begin{bmatrix} w_{1,1} & w_{1,2} & \ldots & w_{1,n} \\ w_{2,1} & w_{2,2} & \ldots & w_{2,n} \\ \ldots & \ldots & \ldots & \ldots \\ w_{n,1} & w_{n,2} & \ldots & w_{n,n} \end{bmatrix} \\
&= \begin{bmatrix} \Gamma_1(x_1) & \Gamma_2(x_1) & \ldots & \Gamma_p(x_1) \\ \Gamma_1(x_2) & \Gamma_2(x_2) & \ldots & \Gamma_p(x_2) \\ \Gamma_1(x_3) & \Gamma_2(x_3) & \ldots & \Gamma_p(x_3) \\ \ldots & \ldots & \ldots & \ldots \\ \Gamma_1(x_n) & \Gamma_2(x_n) & \ldots & \Gamma_p(x_n) \end{bmatrix} \begin{bmatrix} \Gamma_1(x_1) & \Gamma_2(x_1) & \ldots & \Gamma_p(x_1) \\ \Gamma_1(x_2) & \Gamma_2(x_2) & \ldots & \Gamma_p(x_2) \\ \Gamma_1(x_3) & \Gamma_2(x_3) & \ldots & \Gamma_p(x_3) \\ \ldots & \ldots & \ldots & \ldots \\ \Gamma_1(x_n) & \Gamma_2(x_n) & \ldots & \Gamma_p(x_n) \end{bmatrix}^T \\
&= \sum_s \Gamma_s(x_i)\Gamma_s(x_j)
\end{aligned}
\tag{10.36}
$$

We remember that w_{ij} is the connection element or self associative memory.

$$
Aw = \begin{bmatrix} F(x_1) \\ F(x_2) \\ F(x_3) \\ \cdots \\ F(x_n) \end{bmatrix} = \begin{bmatrix} \Gamma_1(x_1) & \Gamma_2(x_1) & \cdots & \Gamma_p(x_1) \\ \Gamma_1(x_2) & \Gamma_2(x_2) & \cdots & \Gamma_p(x_2) \\ \Gamma_1(x_3) & \Gamma_2(x_3) & \cdots & \Gamma_p(x_3) \\ \cdots & \cdots & \cdots & \cdots \\ \Gamma_1(x_n) & \Gamma_2(x_n) & \cdots & \Gamma_p(x_n) \end{bmatrix} \begin{bmatrix} w_1 \\ w_2 \\ \cdots \\ w_p \end{bmatrix}
$$

$$
= w_1 \begin{bmatrix} \Gamma_1(x_1) \\ \Gamma_1(x_2) \\ \Gamma_1(x_3) \\ \cdots \\ \Gamma_1(x_n) \end{bmatrix} + w_2 \begin{bmatrix} \Gamma_2(x_1) \\ \Gamma_2(x_2) \\ \Gamma_2(x_3) \\ \cdots \\ \Gamma_2(x_n) \end{bmatrix} + \cdots + w_p \begin{bmatrix} \Gamma_n(x_1) \\ \Gamma_n(x_2) \\ \Gamma_n(x_3) \\ \cdots \\ \Gamma_n(x_n) \end{bmatrix}
$$

$$(10.37)$$

For the extension of the associative memory we can enlarge the Hopfield model with a more complex projection operator where X are not orthonormal set of vectors.

$$
w = (A^T A)^{-1} A F
$$
$$
F^* = A(A^T A)^{-1} A F = W F = Q F
$$

$$(10.38)$$

In conclusion, the novel Hopfield–Like network has the property to be non Euclidean, memory matrix. Self associative memory is given by general projection operator where the elementary basis vectors X are dependent on each other and there is the correlation among the elementary basis fields.

10.7   Kohonen Self Organizing Maps by Morphogenetic Computing [1–3]

Given the set of points with three dimensions vectors of weights w_{ij} in Fig. 10.25.

    The basic principle of the Self-Organizing Map is to adjust these weight vectors until the map represents a picture of the input data set. The goal of learning in the Self-Organizing Map is to cause different parts of the network to respond similarly to certain input patterns. Figure 10.26 shows the principle of input and output in Self-Organizing Map.

    In morphogenetic computing and one step method the similarity between input vectors X and weights vectors W can be obtained by the projection operator Q.

$$
QW = X(X^T X)^{-1} X^T W
$$

$$(10.39)$$

For the orthogonality of the projection operator the QW vectors of weights is at the minimum of the projection space of the inputs.

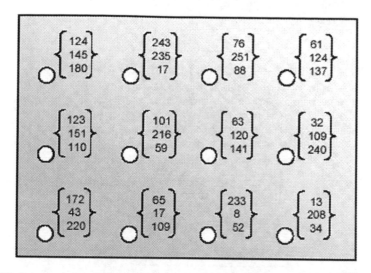

Fig. 10.25 Set of points in Kohonen self organise map with three dimensions weigths

Example 10.3 Given the input vectors

$$x_1 = \begin{bmatrix} 1 \\ 1 \\ 0 \\ 0 \end{bmatrix}, x_2 = \begin{bmatrix} 0 \\ 0 \\ 0 \\ 1 \end{bmatrix}, X = \begin{bmatrix} 1 & 0 \\ 1 & 0 \\ 0 & 0 \\ 0 & 1 \end{bmatrix}$$

$$w_1 = \begin{bmatrix} 0.7 \\ 0.2 \\ 0.5 \\ 0.9 \end{bmatrix}, w_2 = \begin{bmatrix} 0.1 \\ 0.7 \\ 0.3 \\ 0.6 \end{bmatrix}, w_3 = \begin{bmatrix} 0.32 \\ 0.4 \\ 0.3 \\ 0.63 \end{bmatrix}, W = \begin{bmatrix} 0.7 & 0.1 & 0.32 \\ 0.2 & 0.7 & 0.4 \\ 0.5 & 0.3 & 0.3 \\ 0.9 & 0.6 & 0.63 \end{bmatrix}$$

So we have

$$QW = \begin{bmatrix} 1 & 0 \\ 1 & 0 \\ 0 & 0 \\ 0 & 1 \end{bmatrix} \left(\begin{bmatrix} 1 & 0 \\ 1 & 0 \\ 0 & 0 \\ 0 & 1 \end{bmatrix}^T \begin{bmatrix} 1 & 0 \\ 1 & 0 \\ 0 & 0 \\ 0 & 1 \end{bmatrix} \right)^{-1} \begin{bmatrix} 1 & 0 \\ 1 & 0 \\ 0 & 0 \\ 0 & 1 \end{bmatrix}^T \begin{bmatrix} 0.7 & 0.1 & 0.32 \\ 0.2 & 0.7 & 0.4 \\ 0.5 & 0.3 & 0.3 \\ 0.9 & 0.6 & 0.63 \end{bmatrix}$$

$$= \begin{bmatrix} 0.45 & 0.4 & 0.36 \\ 0.49 & 0.4 & 0.36 \\ 0 & 0 & 0 \\ 0.9 & 0.6 & 0.63 \end{bmatrix}$$

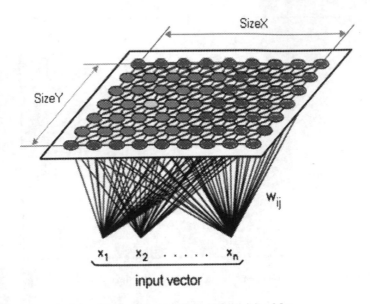

Fig. 10.26 The principle of input and output in Self-Organizing Map

The weights

$$w_{ij} = \begin{bmatrix} 0.45 & 0.4 & 0.36 \\ 0.49 & 0.4 & 0.36 \\ 0 & 0 & 0 \\ 0.9 & 0.6 & 0.63 \end{bmatrix}$$

are the most similar weights to the input data given the initial random values.

Example 10.4 The following is the canonical form by projection operator as a particular case of Kohonen network modeled by morphogenetic computing.
   Given the input vectors

$$X = \begin{bmatrix} 1 & 1 \\ 1 & -1 \\ -1 & 1 \\ -1 & -1 \end{bmatrix} \text{ and the weights } W = \begin{bmatrix} 1 & 1 \\ 1 & -1 \\ -1 & 1 \\ -1 & -1 \end{bmatrix} + \begin{bmatrix} 0.5 & 1 \\ 0.5 & 1 \\ 0.5 & 1 \\ 0.5 & 1 \end{bmatrix} = X + D$$

We have the new weights

$$QW = \begin{bmatrix} 1 & 1 \\ -1 & 1 \\ 1 & -1 \\ -1 & -1 \end{bmatrix} \left(\begin{bmatrix} 1 & 1 \\ -1 & 1 \\ 1 & -1 \\ -1 & -1 \end{bmatrix}^T \begin{bmatrix} 1 & 1 \\ -1 & 1 \\ 1 & -1 \\ -1 & -1 \end{bmatrix} \right)^{-1} \begin{bmatrix} 1 & 1 \\ -1 & 1 \\ 1 & -1 \\ -1 & -1 \end{bmatrix}^T$$

$$\left(\begin{bmatrix} 1 & 1 \\ 1 & -1 \\ -1 & 1 \\ -1 & -1 \end{bmatrix} + \begin{bmatrix} 0.5 & 1 \\ 0.5 & 1 \\ 0.5 & 1 \\ 0.5 & 1 \end{bmatrix} \right) = \begin{bmatrix} 1 & 1 \\ -1 & 1 \\ 1 & -1 \\ -1 & -1 \end{bmatrix}$$

The projection operator changes the weights in a way to eliminate the given translation to be equal to input vectors.

In Fig. 10.27, the input vectors are the black points symmetric to the origin. The diamond points are non-symmetric and the Kohonen neurons that assume the same black point position after the projection operator. Now we rotate and translate W to get Fig. 10.28. The non-symmetric point or neurons assume the same property (symmetry) of the input vectors but are not equal (similar) to the input symmetric vectors or points.

After the projection, QW has the same symmetry of the input vector. In conclusion, the projection operator is like the self organizing maps that preserves the form or properties (symmetry) of the input vector. The similarity means the same morphology in the morphogenetic computing.

Fig. 10.27 The position before rotation of W

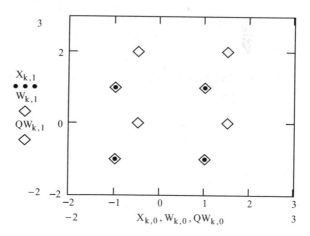

Fig. 10.28 The position after rotation of W

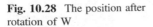

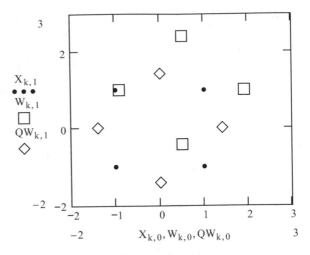

10.8 Morphogenetic Computing Learning with Noising to Learn Patterns and Retrieve Patterns

In many cases we have to compute the weights that are orthogonal to the given input data. In fact for neural associative memory the goal is to design a neural network capable of memorizing a large set of patterns from a data set X (learning phase), and recalling them later in presence of noise (recall phase). Each pattern $x = \{x_1, x_2, \ldots, x_n\}$ is a vector of length n or a field, our focus is to memorize the patterns with strong local correlation among the entries. More specially, we divide the entries of each pattern x into L overlapping subpatterns due to overlaps, a pattern node can be a member of multiple subpatterns. We denote the ith subpattern by $x^i = \left\{ x_1^i, x_2^i, \ldots, x_{n_i}^i \right\}$ the Fig. 10.23 gives the example of the pattern and sub pattern (Fig. 10.29).

The learning process is given by the weight matrix W for which we have the orthogonal condition (10.40).

$$W^i x^i = 0 \qquad (10.40)$$

The weights $w_{i,j}$ is the dual space of the subpatterns system. Now with the projection operator it is possible to compute these weights with only one step as shown in (10.41).

$$W = w_{i,j} = I - X[X^T X]^{-1} X^T \qquad (10.41)$$

In fact we have (10.42)

Fig. 10.29 Any box is sub pattern of the wave interference figure

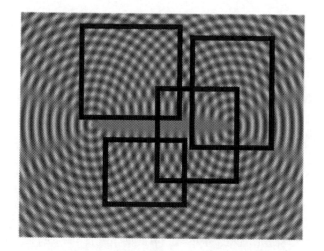

$$WX = (I - X[X^T X]^{-1} X^T)X = 0 \qquad (10.42)$$

Example 10.5 Given the pattern $x = \begin{bmatrix} 0 \\ 1 \\ 1 \\ 0 \end{bmatrix}$. It includes two sub patterns

$x^1 = \begin{bmatrix} 0 \\ 1 \\ 0 \\ 0 \end{bmatrix}, x^2 = \begin{bmatrix} 0 \\ 0 \\ 1 \\ 0 \end{bmatrix}$.

Now the weights are

$$W = \begin{bmatrix} 1 & 0 & 0 & 0 \\ 0 & 1 & 0 & 0 \\ 0 & 0 & 1 & 0 \\ 0 & 0 & 0 & 1 \end{bmatrix} - \begin{bmatrix} 0 & 0 \\ 1 & 0 \\ 0 & 1 \\ 0 & 0 \end{bmatrix} \left(\begin{bmatrix} 0 & 0 \\ 1 & 0 \\ 0 & 1 \\ 0 & 0 \end{bmatrix}^T \begin{bmatrix} 0 & 0 \\ 1 & 0 \\ 0 & 1 \\ 0 & 0 \end{bmatrix} \right)^{-1} \begin{bmatrix} 0 & 0 \\ 1 & 0 \\ 0 & 1 \\ 0 & 0 \end{bmatrix}^T$$

$$= \begin{bmatrix} 1 & 0 & 0 & 0 \\ 0 & 0 & 0 & 0 \\ 0 & 0 & 0 & 0 \\ 0 & 0 & 0 & 1 \end{bmatrix}$$

To retrieve the sub pattern from the weights W, we compute the inverse process.

$$X = \begin{bmatrix} 1 & 0 & 0 & 0 \\ 0 & 1 & 0 & 0 \\ 0 & 0 & 1 & 0 \\ 0 & 0 & 0 & 1 \end{bmatrix} - \begin{bmatrix} 1 & 0 \\ 0 & 0 \\ 0 & 0 \\ 0 & 1 \end{bmatrix} \left(\begin{bmatrix} 1 & 0 \\ 0 & 0 \\ 0 & 0 \\ 0 & 1 \end{bmatrix}^T \begin{bmatrix} 1 & 0 \\ 0 & 0 \\ 0 & 0 \\ 0 & 1 \end{bmatrix} \right)^{-1} \begin{bmatrix} 1 & 0 \\ 0 & 0 \\ 0 & 0 \\ 0 & 1 \end{bmatrix}^T = \begin{bmatrix} 0 & 0 \\ 1 & 0 \\ 0 & 1 \\ 0 & 0 \end{bmatrix}$$

Now when we have a pattern with noise given by expression (10.43)

$$y^i = x^i + \varepsilon^i \tag{10.43}$$

We can detect the noise because we have (10.44)

$$Wy = Wx + W\varepsilon^i = W\varepsilon^i \tag{10.44}$$

So we can clean the sub-partner form its noise and retrieve the original pattern x. The task of a neural associative memory is to retrieve a set of previously memorized patterns from their noisy versions by using a network of neurons. We show that with the projection operator we can easily solve the previous problems that are similar to the Convolutional Neural Associative Memories.

10.9 Conclusion

In this chapter we show that supervised and unsupervised neural network can be computed by a new model, projection method, that takes care of the invariance or rules in the samples as parts of special universe. In traditional physical system rules are modeled by symbolic differential equations that we solve to find the behavior of elements in the physical universe. Now we know that these differential equations are very abstract and difficult to find the wanted behavior, and the brain cannot use differential equations or symbolic expressions as we know in traditional mathematical sense. The brain uses samples that include the universal rules in the implicit way. With samples we can build associative memory with traditional orthonormal property or propagator that extends the traditional associative memory. Now the brain takes sensor input information and associative internal structure that changes the sensor information in a way to have internal data that satisfy the rules which are previously learned by samples of input and output (associative memory or memory with rules). So we think that for digital memory in the computer and associative memory the former is a passive memory and the latter is an active memory that changes the data in a way to include the rules of the external universe or environment in an implicit way. With extension of the associative memory we include Hopfield neural network, Kohonen self-organizing maps, pattern recognition and supervised neuron. For supervised neuron we create a new algorithm by which it is possible to avoid most of the problems of back propagation. The difference between the brain and universal physics is that rules in the physical part are fixed and unable to be changed, in the brain there are also rules as in the physical universe but we can control the rules to obtain our aims by projection method that solve possible contradiction in the implementation of the rules. Solutions is evident in the brain as expansion with new neurons or by new type of neural network parameters. We see that we can have different levels of solutions. The first solve completely all possible contradictions with a maximum number of neurons, the second solve the contradiction but with a bias or threshold, the third solve again the contradiction by threshold but the instability of the system increase. From the first to the last levels we solve always the contradiction with a reduction of the neurons numbers but when the number of the neurons decrease we obtain solutions of the same problem but we increase the instability so the neural network became more and more sensible to the noise.

References

1. K. Deb, S. Agrawal, A. Pratap, T. Meyarivan, A fast elitist non-dominated sorting genetic algorithm for multi-objective optimization: NSGA-II, in *Parallel Problem Solving from Nature PPSN VI Lecture Notes in Computer Science*, vol. 2000, pp. 849–858 (1917)
2. W.J. Freeman, *Mass Action in The Nervous System* (Academic Press, New York San Francisco London, 1975)
3. S. Haykin, *Neural Networks, A Comprehensive Foundation* (Prentice-Hall, Englewood Cliffs, 1999)
4. J.H. Holland, *Adaptation in Natural and Artificial Systems* (The University of Michigan Press, Ann Arbor, 1975)
5. Y.H. Hu, Associative Learning and Principal Component Analysis, Lecture 6 Notes (2003)
6. T. Kohonen, The self-organizing map. Proc. IEEE **78**(9), 1464–1480 (1990)
7. T. Kohonen, Engineering applications of the self-organizing map. Proc. IEEE **84**(10), 1358–1384 (1996)
8. T. Kohonen, *Self-Organizing Maps* (Springer, Berlin, 2001)
9. J.R. Koza, Genetic evolution and co-evolution of computer programs
10. R.K.W.J. Kozma Freeman, Intermittent spatial—temporal desynchronization and sequenced synchrony in ECoG signal. Interdiscipl. J. Chaos **18**, 037131 (2008)
11. R.P. Lippmann, An introduction to computing with neural nets. IEEE Trans. Acoust, Speech. Signal. Process.ASSP. **4**, 4–22 (1987)
12. F. Marcelloni, G. Resconi, P. Ducange, Morphogenetic approach to system identification. Int. J. Intell. Syst. **24**(9), 955–975 (2009)
13 R. McEliece et al., The capacity of Hopfield Associative Memory. Trans. Inf. Theory **1**, 33–45 (1987)
14. C. Mead, Analog VLSI and neural systems (1989)
15. C. Mead, Neuromoprhic electronic systems. Proc. IEEE **78**(10) (1990)
16. M. Nikravesh, G. Resconi, Morphic computing: morphic systems and morphic system of systems (M-SoS. Appl. Comput. Math. **9**(1), 85–94 (2010)
17. K.H. Pribram, *Languages of the Brain: Experimental Paradoxes and Principles in Neuropsychology* (Brandon House, New York, 1971)
18. K.H. Pribram (ed.), *Brain and Perception: Holonomy and Structure in Figural Processing* (Lawrence Erlbaum Associates, Hillsdale, New Jersey, 1991)
19. J.C.F. Pujol, R. Poli, Evolving the topology and the weights of neural networks using a dual representation
20. G. Resconi, The morphogenetic neuron. Comput. Intell. Soft Comput. Fuzzy-Neuro Integr. Appl. **162**, 304–332 (1998)
21. G. Resconi, Modelling fuzzy cognitive map by electrical and chemical equivalent circuits Joint, in *Conference Information Science*, Salt lake City Center USA, 8–24 July 2007
22. G. Resconi, Inferential process by morphogenetic system. Comput. Intell. Decis. Control. 1 (48), 85–90 (2008).

© Springer International Publishing AG 2017

G. Resconi et al., *Introduction to Morphogenetic Computing*,

Studies in Computational Intelligence 703, DOI 10.1007/978-3-319-57615-2

23. Resconi, G, Morphogenetic evolution, in *2013 International Conference on System Science and Engineering (ICSSE)*, pp. 357–362

24. G. Resconi, Conflict compensation, redundancy and similarity in data bases federetion, in *Transaction on Computational Collective Intelligence XIV*, ed. by N.T. Nguyen, 16 August 2014 (Springer, Berlin, 2014)

25. G. Resconi, R. Kozma, *Geometry image of neurodynamics* (NCTA, 2012)

26. G. Resconi, M. Nikravesh, Morphic computing. Appl. Soft Comput. **8**(3), 1164–1177 (2008)

27. G. Resconi, V.P. Srini, Electrical circuit as a morphogenetic system. GEST Int. Trans. Comput. Sci. Eng. **53**(1), 47–92 (2009)@@

28. G. Resconi, A.J. van der Wal, Morphogenic neural networks encode abstract rules by data. Inf. Sci. **142**(1–4), 249–273 (2002)

29. J. Rinzel, B. Ermentrout, Analysis of neural excitability and oscillations, in *Methods in Neural Modeling*, ed. by C.Koch, I.Segev (MIT Press, 1998), pp. 251–291

30 G.X. Ritter, P. Sussner, An introduction to morphological neural networks, in *Proceedings of the 13th International Conference on Pattern Recognition*, Vienna, Austria, pp. 709–711 (1996)

31. G.X. Ritter, P. Sussner, J.L. Diaz de Leon, Morphological associative memories. IEEE Trans. Neural Netw. **9**(2), 281–293 (1998)

32. E. Rowe, *The Dynamical Systems Model of the Simnple Genetic Algorithm, Theoretical Aspects of Evolutionary Computing* (Springer, Berlin, 2001)

33. G.S. Snider 2008 Hewlett packard Laboratory, in *Berkeley Conference on Memristors*

34. A.B. Torralba, Analogue architectures for vision cellular neural networks and neuromorphic circuits, Doctorat thesis, Institute National Polytechnique Grenoble, Laboratory of Images and Signals (1999)

35. T. Vidal, T.G. Crainic, M. Gendreau, N. Lahrichi, W. Rei, A hybrid genetic algorithm for multidepot and periodic vehicle routing problems

Printed in the United States
By Bookmasters